Sustainable Buildings and Cities (BVD'2017)

IOP Conference Series: Materials Science and Engineering
Volume 353

Fez, Morocco
6 – 7 December 2017

ISBN: 978-1-5108-6311-8
ISSN: 1757-8981

Printed from e-media with permission by:

Curran Associates, Inc.
57 Morehouse Lane
Red Hook, NY 12571

Some format issues inherent in the e-media version may also appear in this print version.

Printed by Curran Associates, Inc. (2018)

For permission requests, please contact the Institute of Physics
at the address below.

Institute of Physics
Dirac House, Temple Back
Bristol BS1 6BE UK

Phone: 44 1 17 929 7481
Fax: 44 1 17 920 0979

techtracking@iop.org

Additional copies of this publication are available from:

Curran Associates, Inc.
57 Morehouse Lane
Red Hook, NY 12571 USA
Phone: 845-758-0400
Fax: 845-758-2633
Email: curran@proceedings.com
Web: www.proceedings.com

Table of contents

Volume 353

Sustainable Buildings and Cities

6–7 December 2017, Fez Meknes, Morocco

Accepted papers received: 18 April 2018
Published online: 16 May 2018

Preface

Papers

IOP Conf. Series: Materials Science and Engineering **353** (2018) 011001 doi:10.1088/1757-899X/353/1/011001

Preface
Sustainable Buildings and cities'17

Overview

The Private University of Fez is aware that its development is dependent on the actions of partnerships and collaborations in the fields of training, research, innovation and expertise.

It is a policy that has been strengthened and supported within the UPF, which, while seeking to diversify and multiply its partnerships around the "win-win" logic, is focusing on making them productive and functional.

It is in this context that subscribes the co-organization of this international congress by the private University of Fez and University Moulay Ismail.

The theme "Sustainable buildings and cities" that we have chosen for the research topics of this congress is one of the first sectors of economic activities in Morocco, representing almost 6% of national PIB. It is also one of the most energy-consuming sectors and one of the largest consumers of natural resources. It is therefore a major issue in the Energetical and Ecological Transition of our country.

For several years, ambitious objectives have been set by the public authorities to develop the renewable energies' part, to put in place a plan for energy efficiency and to develop a collective awareness towards the environment.

It is certain that these ambitious objectives can only be achieved through a collaboration between all the people involved in the sector, which necessarily leads to a transversality of the analyses between economic and social sciences and those of the engineer.

The interdisciplinary of the theme "Sustainable buildings and cities" will bring together different academic and professional stakeholders interested not only in the scientific and technical aspects of new developments in the domain, but also to the regulatory, economic and social aspects necessary for the implementation and application of good practices.

It was around all these issues that we debated and exchanged during the two days this congress lasted.

BVD'2017 Chairman
Pr. Mohammed Ouazzani Jamil

IOP Conf. Series: Materials Science and Engineering **353** (2018) 011001 doi:10.1088/1757-899X/353/1/011001

Organizers

IOP Conf. Series: Materials Science and Engineering **353** (2018) 011001 doi:10.1088/1757-899X/353/1/011001

Photographs

IOP Conf. Series: Materials Science and Engineering **353** (2018) 011001 doi:10.1088/1757-899X/353/1/011001

IOP Conf. Series: Materials Science and Engineering **353** (2018) 011001 doi:10.1088/1757-899X/353/1/011001

IOP Conf. Series: Materials Science and Engineering **353** (2018) 011001 doi:10.1088/1757-899X/353/1/011001

IOP Conf. Series: Materials Science and Engineering **353** (2018) 011001 doi:10.1088/1757-899X/353/1/011001

giz

IOP Conf. Series: Materials Science and Engineering **353** (2018) 011001 doi:10.1088/1757-899X/353/1/011001

UPF

& 07 Décembre 2017

IOP Conf. Series: Materials Science and Engineering 353 (2018) 011001 doi:10.1088/1757-899X/353/1/011001

IOP Conf. Series: Materials Science and Engineering **353** (2018) 011001 doi:10.1088/1757-899X/353/1/011001

Committees

Honour Committee

H. SAHBI	Président	UMI
M.A. LAHLOU	Président	UPF
S. DAOUDI	Recteur	UPF
Y.EL AMRAOUI	ENSAM MEKNES-UMI	
H. ELMIR	FS Meknès- UMI	
A. BOUAYAD	FSJES Meknès- UMI	
O.OUSSOUADDI	Vice Président UMI	
A. EL OUALI	Vice Président UMI	

IOP Conf. Series: Materials Science and Engineering **353** (2018) 011001 doi:10.1088/1757-899X/353/1/011001

General Chairs

Chair
Mohammed OUAZZANI JAMIL
(UPF)

Co- Chair
Ahmed LAAROUSSI
(UPF)

Co- Chair
Mohammed KEROUAD FS Meknès
(UMI)

IOP Conf. Series: Materials Science and Engineering **353** (2018) 011001 doi:10.1088/1757-899X/353/1/011001

Organization Committee

M. KEROUAD	***FS -UMI***	M.OUAZZANI JAMIL	***UPF***
M.HADDAD	***FS –UMI***	Z. ITAHRIOUAN	***UPF***
T.EL RHAFIKI	***ENSAM –UMI***	T. HAJJI	***UPF***
A. ALAMI HASSANI	***UPF***	I.AKHRIF	***UPF***
S. FAQUIR	***UPF***	A. GAGA	***UPF***
M. EL BOUMASHOULI	***UPF***	M. ELALAOUI	***UPF***
F. HARCHLI	***UPF***	M. BENABDELLAH	***UPF***
M. CHAABA	***ENSAM –UMI***	K. CHERKAOUI	***ENSAM –UMI***
A. LAAZIZI	***ENSAM –UMI***	M. EL JAI	***EEMGC –UEMF***

Program Committee

M. KEROUAD	***FS MEKNES-UMI***	F. ABDI	***FST FES-USMBA***
Y.EL AMRAOUI	***ENSAM MEKNES-UMI***	F.ZERROUK	***EST FES-USMBA***
M.ASBIK	***ENSET -RABAT -UM5***	P. HUOVILA	***VTT- Finland***
M.HADDAD	***FS MEKNES-UMI***	A.EL ARRIM	***FST TANGER-UAE***
A.AINANE	***FS MEKNES-UMI***	S.PINCEMIN	***EPF Montpelier- France***
L.BEJJIT	***EST MEKNES-UMI***	F. STEPHAN	***EPF Montpelier- France***
A. LAAZIZI	***ENSAM MEKNES-UMI***	K.LAFDI	***University of Dayton- USA***
A. BOUATEM	***ENSAM MEKNES-UMI***	A.BENYOUSSEF	***Académie HassanII-MAScIR***
K. CHERKAOUI	***ENSAM MEKNES-UMI***	Y. BOUGHALEB	***UCD- EL JADIDA***
H. AHRCHOUI	***FS MEKNES-UMI***	M.HAMEDOUN	***MAScIR***
M. EL JAI	***EEMGC-UEMF***	M.OUAZZANI JAMIL	***FSI - UPF***
A.REZZOUK	***FS FES -USMBA***	A.LAAROUSSI	***ESMAB- UPF***
J.KHARBACH	***FS FES -USMBA***	M. EL BOUMASHOULI	***FSI - UPF***
I.ZORKANI	***FS FES -USMBA***	I.AKHRIF	***FSI - UPF***
A.SALI	***FS FES -USMBA***	A. ALAMI HASSANI	***FSI - UPF***
M. ELQANDIL	***FS FES -USMBA***	A. GAGA	***FSI - UPF***
M. MHARZI	***FS FES -USMBA***	T. HAJJI	***FSI - UPF***
A.HOURMATALLAH	***ENS FES -USMBA***	Z. ITAHRIOUAN	***FSI - UPF***
R.JABRANE	***FST FES-USMBA***	S. FAQUIR	***FSI - UPF***
A.AHAITOUF	***FST FES-USMBA***	M. ELALAOUI	***FSI - UPF***
N.ES-SBAI	***FST FES-USMBA***	M. BENABDELLAH	***FSI - UPF***
A.MECHAQRANE	***FST FES-USMBA***	F. HARCHLI	***FSI - UPF***

Students Committee

EL ALLALI IMANE	JAMOUR MOUNTASIR BILAH
AOUINTI OUISSAL	HADI ALAOUI OUAFA
EL YANDOUZI MARWANE	CHARIA KENZA
MOSAMEH NAIMA	BENJAJ ABDERRAHIM
MAMADOU DEMBELE	MARTAH MOHAMMED
BOUDANE MOHAMMED	EL LAGHZZIOUI ISMAIL WASSIM
STEPHANIE BERE	BEN ABDELJALI YASSINE
LAHMER ISMAIL	CHIG HAJAR
DAOUIRI ACHRAF	CHERGUI HOSSAIN
SAMAA YOUSSEF	

IOP Conf. Series: Materials Science and Engineering **353** (2018) 011001 doi:10.1088/1757-899X/353/1/011001

Sponsoring organisations

Deutsche Gesellschaftfür Internationale Zusammenarbeit

Banque Populaire Fès Maroc

Commune Urbaine de Fès

IOP Conf. Series: Materials Science and Engineering **353** (2018) 011001 doi:10.1088/1757-899X/353/1/011001

Keynotes & Invited speakers

Conference title :
Fez, Smart City

Dr Idriss EL AZAMI EL IDRISSI
Président de la Commune de Fès- Maire de la Ville de Fès

Dr. El Azami el Idrissi is president of the municipality of Fez and Mayor of the city of Fez. In 1989, he obtained a bachelor's degree in chemistry from the Faculty of Science at Sidi Mohammed Ben Abdellah University in Fez. In 1993, he obtained a doctorate degree in Applied Chemistry from the University of Poitiers in France. In 1997, He graduated from the Graduate School of National Administration (l'Ecole Nationale d'Administration) in Rabat in Morocco. In 2001, he graduated in economic studies from the Faculty of Legal, Economic and social Sciences from Mohammed V-Agdal University in Rabat. Dr. El Azami El Idrissi holds also a Diploma of Graduate Studies of the Institute of Business Administration of the University of Paris I Panthéon-Sorbonne in Paris (2006) and a Diploma of Financial, economic and banking Studies of Marseille in Marseille in France. In January 2012, he was appointed as a Delegated Minister to the Minister of Economy and Finance in charge of the Budget. Since September 2015, Dr. El Azami El Idrissi has been elected president of the commune of Fez.

Conference title :
Hydrogen Economy For Clean Cities

Pr. Abdelilah BENYOUSSEF
Académie Hassan II des Sciences et Techniques, Maroc

« Received his Ph.D from Paris-Sud University in 1983. He is a permanent member of Hassan II Moroccan Academy of Science and technology since 2006. He is the national coordinator of the axe of skills of condensed material and systems' modeling. He is also the editor-in-chief of the Moroccan Journal of condensed Material. He is the president of the Moroccan Society of Statistical Physics and condensed material. The main topics of interest of Dr. Benyoussef are the ab initio calculation and the Monte Carlo method in the modeling and simulation of new materials for renewable energies; Magnetism and phase transition in condensed material; Complex systems and critical self-organization in statistical physics. He is the co-author of over 400 research publications and book chapters and about 100 conference presentations, including numerous guest articles and conferences invited speaker. He co-chaired or co-hosted several international conferences. He holds a number of patents and has supervised 40 candidates for postgraduate research »

IOP Conf. Series: Materials Science and Engineering **353** (2018) 011001 doi:10.1088/1757-899X/353/1/011001

Conference title :

The Private-Public Partnership. Governance Improvement Instrument for Smart City. Case of the Delegated Management of Water and Electricity Distribution Services and Liquid Sanitation in Rabat-Sale

Dr. Monsef BERBICHE

Centre de Recherche pour l'Environnement, l'Energie et les Déchets- VEOLIA Maroc

« Received his PhD in 1990 from the University of Paul Sabatier, Toulouse, in Electro-technical, automatic and industrial computing on the transient stability of electrical power networks with application on the national Production and transport network of the ONEE. After teaching in the master of electronic, electro-technic and automatic at the University Paul Sabatier in Toulouse, he returned to Morocco where he held several functions at the Ministry of the Interior, notably director of the boards of Kenitra (RAK) and Marrakech (Redeema). Between 2006 and 2014, Dr. BERBICHE was a director at Veolia Services at L'ENVIRONNEMENT MAROC in Morocco in Rabat and most recently he was appointed as general Director of the research Centre for environment, energy and waste (CREED), with VEOLIA SERVICES at L'ENVIRONNEMENT MAROC in Rabat»

Conference title :

Panorama on Self-consumption, from Building to Sustainable City

Dr. Sandrine PINCEMIN

EPF école d'ingénieur(e)s- France

«Dr. Pincemin is an Assistant Professor and researcher at the EPU (l'Ecole Polytechnique Universitaire) of Marseille in France (know before by IUSTI), School of Engineering, she is responsible for the energy branch and the orientation of conventional and sustainable Energies (ECD). In 2002, she obtained a mechanical engineering degree from the EPF. In 2007, she got her Ph.D in Materials and solar energy processes from the PROMES Laboratory of Processes, Materials and Solar Energy at the Perpignan University in France in the domain of High energy Performance for composite materials' energy storage. She worked after that from 2007 until 2012 as studies and research engineer at CSTB Renewable Energies axe in the field of energy storage (thermal and electric) for low consumption and positive energy buildings in connection with the "Grenelle de l'environment". Dr. Pincemin has been working since 2012 at the EPF (old-Ecole Polytechnique feminine), campus of Montpellier where she is co-responsible for the major energy environment. She carries out her research at the Institut d'Électronique du Sud (UMR CNRS 5214) at University of Montpellier. Her research areas consist of the production of electricity by concentrated photovoltaic solar cells, the Mediterranean building (needs, the cold in the Mediterranean region, and the thermal/electrical modeling of the Mediterranean building and optimization) and auto consumption»

IOP Conf. Series: Materials Science and Engineering **353** (2018) 011001 doi:10.1088/1757-899X/353/1/011001

Conference title :
Photovoltaic Concentration: Principle and Engineering

Pr. Ali HAIATOUF
Faculty of Sciences and Techniques (FST), Sidi Mohammed Ben Abdellah University, Fez- Morocco

« Pr. Ahaitouf is a professor and researcher of high education at the Faculty of Sciences and Techniques (FST), Sidi Mohammed Ben Abdellah University of Fez in Morocco. He obtained his PhD in electronics in 1998 and his doctorate degree from the University of Metz in June 1992. He is a director of Renewable Energies and Intelligent Systems' Laboratory (LERSI) and coordinator for the Microelectronics and embedded systems team. His researches cover electronic and microelectronic components, the design of analog and digital integrated circuits as well as multi-junction solar photovoltaic cells under concentration (CPV). He works in collaboration with several international laboratories including the International Joint Unit Georgia Tech Lorraine, UMI2958 GT-CNRS where he held a full researcher position from September 2011 to August 2012 and a chair of excellence on CPV in 2014. He has written and rewritten more than forty articles in international journals. He has also conducted more than a dozen of doctoral dissertations including cotutelle dissertations»

Conference title :
Smart City - Intelligent Mobility

Mr. Mohammed Mouhcine MAAROUFI
Université Privée de Fès & GRUPO TECOPY-Morocco

«Dr. Maaroufi has an engineering diploma from École Hassania des Travaux Publics of Casablanca in Morocco as sell as a graduate degree (Diplôme d'Etudes Approfondies DEA) in civil engineering from the National School of Public Works (l'École National des Travaux Publics of Lyon in France. Dr. Maaroufi was able to perfect his acquirement in project management, by a training on business management at the ISCAE and a master in Project management at the Collège Polytechnique in Paris. He Has a high level of expertise in the management, planning and development of projects as well as in studies, control and audit to ensure continuous improvement. The projects he has dealt with are developed in various types of areas, particularly in the design of transport infrastructures and the establishment of urban mobility master plans. With all of his combined trainings over 20-year experience in the management of infrastructure projects spans in Morocco and also in Mauritania, Senegal, Burkina Faso, Cameroon, Niger, Mali, Côte d'ivoire, Guinea, Chad and Gabon, he was able to master the Sector and more particularly on aspects related to the design, realization and monitoring of major infrastructure projects. Mr. Maaroufi who has held several responsibility positions in the Moroccan Research Office (Bureau d'études Moroccan CID) is currently director and representative in Morocco of the Spanish group GRUPO TECOPY and he is a professor at the Private University of Fez in Morocco and a member of the AMDIC association»

IOP Conf. Series: Materials Science and Engineering **353** (2018) 011001 doi:10.1088/1757-899X/353/1/011001

Conference title:

Energy Performance Assessment of Passive and Semi-Passive Techniques in Buildings in Semi-Arid Climate

Pr Brahim BENHAMOU
Faculty of Science, Cadi Ayyad University, Marrakech-Morocco

« Professor Benhamou holds a PhD in mechanical engineering (thermo-fluid) from the University of Sherbrooke, Canada. He is currently a professor in the Department of Physics at the Faculty of Science, Cadi Ayyad University in Marrakech (UCAM) in Morocco. He is responsible for the "Laboratory of renewable energies and energy efficiency" (EnR2E) at the National Centre for Research & Studies on water and Energy (CNEREE) affiliated with the Cadi Ayyad University in Marrakech. He is also a Local Coordinator of the Pole of Competences Energy at UCAM. He is also the General Secretary of the Moroccan Association of Thermal Sciences. His research focuses on applications of heat and mass transfer (HMT), energy efficiency in buildings (EEB) and thermal solar energy (desalination and CSP plants). He is the coordinator or has participated in several national and international projects funded on the applications of thermal solar energy and energy efficiency in buildings. One of the funded projects led by Pr. Benhamou, still underway, is entitled "Passive systems for air refreshment in the buildings of the climate zone of Marrakech ". He is the co-author of 29 reading committee journals and more than 100 conference papers»

Conference title :

Eco-Construction and Sustainable Development

Mr. Mohammed EL ANBASSI
Architect at DPLG

« Mr. El Anbassiis a DPLG Architect (Graduated By The Government). He has also a postgraduate Diploma in installation of real estate operations and he was trained in accessibility by the ARVHA and the French Ministry of Culture and in the passive building in Paris in France. He has several accomplishments related to development plans, development studies and architectural projects in reference to sustainable development criteria, development of the territory and the accomplishment of architecture projects related to sustainable development within the framework of INDH, and to the master plan of planning and urban development of peripheral municipalities of the agglomeration of Marrakech and 30 communes around Marrakech»

IOP Conf. Series: Materials Science and Engineering **353** (2018) 011001 doi:10.1088/1757-899X/353/1/011001

Conference title :
Eco-Construction and Sustainable Development

Mr. Rachid BOUQARTACHA
Architect at DPLG

« Mr. Bouqartacha is a DPLG architect (Graduated By The Government), ex head of Technical Studies and processes Department at the Technical Affairs Directorate of the Ministry of Housing (Direction des Affaires Techniques du Ministère de l'Habitat), and ex head of the Institute of Construction in local materials (l'Institut de la construction en matériaux locaux « Instituterre ») "Institut-Terre" in Marrakech. He taught "local materials and Architecture" and "Eco-construction and sustainable architecture" at the National Schools of Architecture in Rabat and Marrakech. Mr. Bouqartacha is a winner of the 2nd Prize of building innovation on behalf of the l'Institut-Terre at the 7th International Exhibition of Building 98 in Casablanca. He is also author and co-author of several rehabilitation and construction in local materials or conventional materials project studies. Mr. Bouqartacha is a consultant at the international scale in the field of building using local materials, member of the Regional Council of Architects of Tensift (1994-2006), Vice-President of the Regional Council of Architects of Tensift (2004-2006), member of the National Council of Architects order (2008-2011), and president of the Association "Eco Construction and sustainable Architecture »

Conference title :
Uncertainty of Measurement and Evaluation of the Energy Performance of a Building

Pr. Abdérafi CHARKI
Institut des Sciences et Techniques de l'Ingénieur d'Angers-France

«Pr. Charki has been a professor at ISTA (Institut des Sciences et Techniques de l'Ingénieur d'Angers) in France since 2004. He carries out his research work within the Laris Laboratory (Angevin Laboratory for Research in Systems Engineering) at Angers University in France. His work focuses on estimating measurement uncertainties and evaluating the systems reliability. Expert in metrology and quality, he is involved in several European projects such as AFREQEN (an Erasmus+ European project) to contribute to the development of access to energy, energy independence, the promotion of renewable energies and to the preservation of the environment in North and West Africa by strengthening the skills of professionals working in the energy and environment sectors. He is also involved in the project of the National Research Agency (ANR) OMEGA (Outils MEthodologiques pour la GArantie de performance énergétique) on the aspect "Cities and sustainable buildings"»

IOP Conf. Series: Materials Science and Engineering **353** (2018) 011001 doi:10.1088/1757-899X/353/1/011001

Conference title :
Compressible Soils and Reinforcement Solutions by Rigid or Flexible Inclusions

Mr .Tarik ELMALKI
SEMARTEC, Casablanca- Maroc

«Mr. El Malkigot a double diplomation in 2004 from l'Ecole Hassania des Travaux publics in Morocco and l'Ecole Nationale des Ponts et Chaussées in France. In 2006, he obtained a Master of science in civil Engineering from Sherbrooke University in Québec. He has 13 years of professional experience in the field of geotechnical studies. After working for the CID and ICE/SEPIA, he joined Semartec as a project manager and responsible for the geotechnical activity of Semartec in Morocco. He has gained real expertise through high-value-added projects where knowledge of the field and the master of geotechnical calculation tools go together such as the LGV Kenitra Tangier project whether in Art works or Earth works. He takes also action as an expert on the analysis of constructive conditions, design orientations on projects presenting a predominant soil-structure interface»

IOP Conf. Series: Materials Science and Engineering **353** (2018) 011002 doi:10.1088/1757-899X/353/1/011002

Peer review statement

All papers published in this volume of *IOP Conference Series: Materials Science and Engineering* have been peer reviewed through processes administered by the proceedings Editors. Reviews were conducted by expert referees to the professional and scientific standards expected of a proceedings journal published by IOP Publishing.

IOP Conf. Series: Materials Science and Engineering **353** (2018) 012001 doi:10.1088/1757-899X/353/1/012001

Heat performance resulting from combined effects of radiation and mixed convection in a rectangular cavity ventilated by injection or suction

K. Ezzaraa[1], A. Bahlaoui[1], I. Arroub[1], A. Raji[2], M. Hasnaoui[3] and M. Naïmi[2]

[1] Sultan Moulay Slimane University, Polydisciplinary Faculty, Department of Physics, Team of Applied Physics and New Technologies (EPANT), B. P. 523, Béni-Mellal 23000, Morocco.

[2] Sultan Moulay Slimane University, Faculty of Sciences and Technologies, Department of Physics, Laboratory of Flows and Transfers Modelling (LAMET), B. P. 523, Béni-Mellal 23000, Morocco.

[3] Cadi Ayyad University, Fac ulty of Sciences Semlalia, Department of Physics, Laboratory of Fluid Mechanics and Energetics (LMFE), B.P. 2390, Marrakech, Morocco.

Abstract. In this work, we investigated numerically heat transfer by mixed convection coupled to thermal radiation in a vented rectangular enclosure uniformly heated from below with a constant heat flux. The fresh fluid is admitted into the cavity by injection or suction, by means of two openings located on the lower part of both right and left vertical sides. Another opening is placed on the middle of the top wall to ensure the ventilation. Air, a radiatively transparent medium, is considered to be the cooling fluid. The inner surfaces, in contact with the fluid, are assumed to be gray, diffuse emitters and reflectors of radiation with identical emissivities. The effects of some pertinent parameters such as the Reynolds number, $300 \leq Re \leq 5000$, and the emissivity of the walls, $0 \leq \varepsilon \leq 0.85$, on flow and temperature patterns as well as on the heat transfer rate within the enclosure are presented for the two ventilation modes (injection and suction). The results indicate that the flow and thermal structures are affected by the thermal radiation for the two modes of imposed flow. However, the suction mode is found to be more favorable to the heat transfer in comparison with the injection one.

1. Introduction

Mixed convection heat transfer in ventilated systems continues to be a fertile area of research, due to the interest of the phenomenon in many technological processes, such as the design of solar collectors, thermal design of buildings, air conditioning and recently the cooling of electronic circuit boards. However, the effect of the radiation, due to its complexity, has been overlooked in the majority of the available studies in spite of its significant contribution to the overall heat transfer. Since moderate temperature differences involve significant radiation effects, the fact of neglecting the latter becomes non-realistic.

To understand the dynamical behaviour of heat and fluid flow due to surface radiation a number of computations have been recently presented on natural convection in closed or opened cavities. Related to the problem of differentially heated cavities, Bouafia et al. [1] performed a numerical study of the interaction of natural convection with thermal radiation in a square cavity filled with air. Detailed analysis showed that the surface radiation reduces the convection heat transfer at the hot wall and increases it on the cold one. Also, the radiation promotes the occurrence of instabilities leading to an early transition to the unsteadiness. When the heating is from below, steady and unsteady analyses of surface radiation and natural convection in a cavity heated from below and cooled from above were carried out by Ridouane et al. [2]. They demonstrated the presence of multiple steady state solutions and they mentioned that the obtained steady state regimes is markedly affected by wall emissivity. In the case of side-vented open cavities, conjugate heat transfer by natural convection and radiation was studied experimentally by Ramesh and Merzkirch [3] where opening is localised on the top. They found that, for cavities with low emissive walls, natural convection was the dominant mode and with high emissive walls, both natural convection and radiation were competitive modes contributing equally to the total heat transfer. In this context, Singh et al. [4] performed a numerical study of the radiation and natural convection interactions in a tilted open cavity. The results showed that the heat transfer increases first and then decreases with decreasing the cavity tilt angle for all Rayleigh numbers and emissivity values.

Unfortunately, studies of mixed convection that involve surface radiation in vented enclosure seem to be relatively few in spite of its practical interest. When the ventilation is made by the inlet and outlet ports, Raji and Hasnaoui [5] numerically investigated the interaction between mixed convection and thermal radiation in a ventilated rectangular enclosure by considering two studied configurations (BT: assisting flows and TB: opposing flows). Results of the study showed that the radiation effect leads to a better homogenization of the temperature inside the cavity. It was also observed that in BT configuration due to the effect of radiation, the average temperature of enclosure increases, while the maximum temperature decreases. However, in the TB configuration, the radiation effect causes lowering of both mean and maximum temperatures. More recently, Bahlaoui et al. [6] carried out a numerical study of mixed convection combined with surface radiation inside a partitioned ventilated cavity. It was shown that the relative height of the adiabatic partition contributes to increase the radiative heat transfer component and leads to reduce the convective heat transfer component at the level of the heated wall.

Based on the available studies mentioned in the above literature review, the case of mixed convection in vented cavities with radiant walls hardly starts to arouse attention in spite of its obvious interest. However, it appears that no work was reported on combined mixed convection and surface radiation inside a rectangular enclosure with multiple ports and ventilated by two different modes (injection and suction). The present research is focused to studying such problem. Hence, the consequence of varying the controlling parameters such as the Reynolds number, the emissivity of the walls and the imposed flow mode, on flow and temperature fields and heat removal rate within the cavity have been investigated and discussed.

2. Problem description and mathematical formulation

The schematic diagram of the geometry considered in the study is presented in Fig. 1. It consists of a ventilated horizontal rectangular cavity having an aspect ratio A = 2 and uniformly heated from below with a constant heat flux. The upper horizontal and right vertical walls are considered perfectly insulated, while the left side of the cavity is cooled with a uniform temperature. The physical system is submitted to an imposed ambient air stream which is admitted by injection (Fig. 1a) or suction (Fig. 1b) through two openings located on the lower part of both right and left vertical walls. A third opening is located on the middle of the upper wall for ventilation purposes. The fluid properties are evaluated at a cold temperature. The inner surfaces, in contact with the fluid, are assumed to be gray, diffuse emitters and reflectors of radiation with identical emissivities. The physical model is two-dimensional and the airflow is assumed to be laminar, incompressible and obeying the Boussinesq approximation. Therefore, using the following dimensionless variables:

IOP Conf. Series: Materials Science and Engineering **353** (2018) 012001 doi:10.1088/1757-899X/353/1/012001

$A = L'/H'$; $B = h'/H'$; $x = x'/H'$; $y = y'/H'$; $u = u'/u'_o$; $v = v'/u'_o$; $t = t'u'_o/H'$; $T = \lambda(T' - T'_C)/q'H'$; $\Psi = \Psi'/u'_oH'$; $\Omega = \Omega'H'/u'_o$; $Pr = \nu/\alpha$; $Ra = g\beta q'H'^4/\alpha\nu\lambda$; $Re = u'_oH'/\nu$

the corresponding set of differential equations, under the commonly assumptions, using the vorticity-stream function formulation can be written in dimensionless form as follows:

$$\frac{\partial \Omega}{\partial t} + u\frac{\partial \Omega}{\partial x} + v\frac{\partial \Omega}{\partial y} = \frac{1}{Re}\left[\frac{\partial^2 \Omega}{\partial x^2} + \frac{\partial^2 \Omega}{\partial y^2}\right] + \frac{Ra}{Re^2\,Pr}\frac{\partial T}{\partial x} \quad (1)$$

$$\frac{\partial T}{\partial t} + u\frac{\partial T}{\partial x} + v\frac{\partial T}{\partial y} = \frac{1}{Re\,Pr}\left[\frac{\partial^2 T}{\partial x^2} + \frac{\partial^2 T}{\partial y^2}\right] \quad (2)$$

$$\frac{\partial^2 \Psi}{\partial x^2} + \frac{\partial^2 \Psi}{\partial y^2} = -\Omega \quad (3)$$

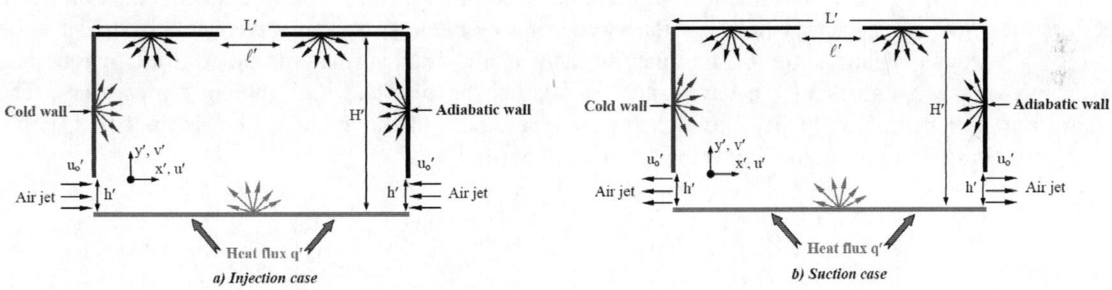

Figure 1. Schematic diagram of the studied configuration: a) Injection case and b) Suction case.

2.1. Boundary conditions

The common boundary conditions, in the dimensionless form, applied to the two ventilation modes can be defined by:

$u = v = 0$	on the rigid walls
$T = 0$	on the left vertical cold wall
$-\frac{\partial T}{\partial y} + N_r Q_r = 1$	on the lower horizontal heated wall
$-\frac{\partial T}{\partial n} + N_r Q_r = 0$	on the adiabatic walls
$\Psi = 0$	on the lower horizontal heated wall

"n" indicates the outward direction normal to the considered adiabatic wall.

The appropriate dimensionless boundary conditions related to the injection or suction cases can be written as:

Injection case :

$T = v = \Omega = 0$, $u = 1$ and $\Psi = y$	at the left inlet of the cavity
$T = v = \Omega = 0$, $u = -1$ and $\Psi = -y$	at the right inlet of the cavity
$\Psi = B$	between the left inlet port and the upper outlet port
$\Psi = -B$	between the right inlet port and the upper outlet port

For this injection mode, the boundary conditions are unknown at the upper outlet opening. Values of u, v, T, Ψ and Ω are extrapolated at each time step by considering zero second derivatives of these variables at the exit of the cavity.

IOP Conf. Series: Materials Science and Engineering **353** (2018) 012001 doi:10.1088/1757-899X/353/1/012001

Suction case :

$T = 0$	at the upper inlet port
$u = -1, v = 0, \Psi = -y$ and $\Omega = 0$	at the left outlet port
$u = 1, v = 0, \Psi = y$ and $\Omega = 0$	at the right outlet port
$\Psi = -B$	between the left outlet port and the upper inlet port
$\Psi = B$	between the right outlet port and the upper inlet port

For this suction mode, the boundary conditions for u, v, Ψ and Ω are unknown at the upper inlet opening whereas the temperature T is unknown at both left and right outlet openings. Similarly to the previous case, values of these variables are obtained at each time step by considering zero second derivatives of these variables at these openings.

2.2. Radiation equations

The enclosure analysis, with radiosity-irradiation formulation for the evaluation of the radiosities of all the walls elements, is used for making the calculation of the radiative heat exchange between the cavity and its surrounding. The walls of the cavity are subdivided into N = 600 elementary segments for the grid retained (201×101). Each segment is sufficiently small to be considered at uniform temperature. The shape factors between the isothermal elementary surfaces were determined by the Hottel's [7] crossed string method. The inner surfaces of the enclosure are assumed to be opaque and diffuse-grey. The openings of the cavity are assumed to be black at the ambient temperature. The general non-dimensional radiosity equation for the ith element of the enclosure, which is filled with a radiatively non-participating medium, like air, may be written as:

$$J_i = \varepsilon_i \left(\frac{T_i}{T_o} + 1 \right)^4 + (1 - \varepsilon_i) \sum_{j=1}^{N} F_{ij} J_j \qquad (4)$$

Where $J_i = J'_i / \sigma T'^4_C$ and $T_o = \lambda T'_C / q' H'$ are the dimensionless radiosity and the dimensionless reference temperature, respectively.

2.3. Heat transfer

The mean Nusselt numbers, characterizing the contributions of mixed convection and thermal radiation through the heated wall, are respectively determined by:

$$Nu_H(cv) = -\frac{1}{A} \int_0^A \frac{1}{T} \left(\frac{\partial T}{\partial y} \right) \bigg|_{y=0} dx \quad ; \quad Nu_H(rd) = \frac{1}{A} \int_0^A \frac{1}{T} (N_r Q_r) \bigg|_{y=0} dx \qquad (5)$$

Where $N_r = \sigma T'^4_C / q'$ and $Q_r = Q'_r / \sigma T'^4_C$ are the convection-radiation interaction parameter and the dimensionless net radiative heat flux, respectively.

The overall Nusselt number, Nu_H, is evaluated as being the sum of the corresponding convective and radiative Nusselt numbers; i.e. $Nu_H = Nu_H(cv) + Nu_H(rd)$.

3. Numerics

The set of coupled partial differential equations, Eqs. (1)-(3), were discretized using a finite differences method. The first and second derivatives of the diffusive terms were approached by central differences method while the advection terms are handled using a second order upwind scheme. Then, an alternating direction implicit (ADI) procedure was used to perform the time integration for the equations (1) and (2). At each time step, the Poisson equation, Eq. (3), was treated by using the point successive over-relaxation method (PSOR) with an optimum over-relaxation coefficient equal to 1.88 for the grid adopted in this study. The set of Eq. (4), representing the radiative heat exchange between the different sub-surfaces of the cavity, was solved by using the Gauss- Seidel method. The numerical computations were started using a pseudo-conductive regime or pure mixed convection solution as initial condition but the final stat is the same steady state solution.

4. Code validation and grid independence study

The numerical code was tested against the results of Wang et al. [8], in the case of a square cavity differentially heated, by simulating the mean total Nusselt numbers evaluated on the hot wall for Ra ranging from 10^4 to 10^6 and ε varying from 0 to 0.85. The maximum deviations observed remain within 1.62 % (Fig. 2). In addition, an overall energy balance for the system was systematically checked for all the computations. Thus, the energy released by the heating wall to the fluid is evaluated and compared with that leaving the cavity through the cold wall and the openings. The maximum relative variations generated are lower than 2 %.

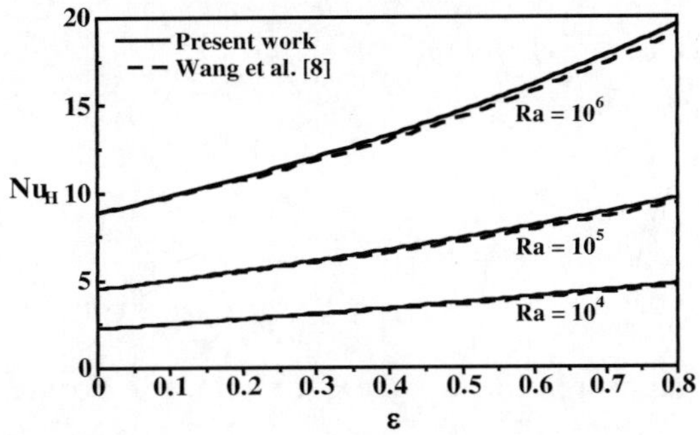

Figure 2. Effect of Ra and ε on the mean total Nusselt number, Nu_H, evaluated on the hot wall of a differentially heated square cavity for $T'_H = 298.5$ K and $T'_C = 288.5$ K.

The calculations were performed with a uniform grid of 201×101 for both x and y directions; grid selected was estimated to be appropriate for the present study since it permits a good compromise between the computational cost and the accuracy required. The results obtained with this grid were comparable to those obtained with a finer grid of 281×141. In fact, the refinement of the grid to 281×141 involves maximum variations lower than 0.86 % and 1.37 % in terms of Ψ_{max} and Nu, respectively (Table 1).

Table 1. Grid sensitivity check on the obtained results for Ra = 5×10^6 (Nr = 2.02) and different values of Re and ε.

	Re = 300				Re = 1500			
	ε = 0.15		ε = 0.85		ε = 0.15		ε = 0.85	
Grids	Nu_H	Ψ_{max}	Nu_H	Ψ_{max}	Nu_H	Ψ_{max}	Nu_H	Ψ_{max}
201×101	15.882	0.231	20.220	0.224	19.101	0.236	24.761	0.233
281×141	16.086	0.231	20.433	0.225	19.362	0.238	25.048	0.235

5. Results and discussion

In the present study, the value of the temperature of the ambient imposed flow was considered constant at $T'_C = 298.15$ K. The fixed value of Rayleigh number (Ra = 5×10^6), retained in this work, induces automatic values of the parameters N_r and T_o equal to 2.02 and 0.704, respectively. Also, the relative height of the vertical openings, h, and the relative length of the upper horizontal opening, λ, are maintained constants at 1/5 and 2/5 respectively.

Typical streamlines and isotherms illustrating radiation effect on the flow structure and temperature patterns, in the injection case, are presented in Figs. 3a-3b for Re = 300 and various values of ε. In the absence of radiation effect (ε = 0), Fig. 3a shows that the forced flow enters horizontally through the two vertical inlet openings and then ascends vertically at the middle of the cavity before leaving the latter through the upper outlet opening. Also, it is seen the existence of two closed cells surmounting the open streamlines. The formation of these cells, located in the corners of the upper part of the cavity, is due to the shear effect. The corresponding isotherms are tightened at the level of the heated bottom wall indicating a good convective heat exchange. In addition, a thin horizontal thermal boundary layer is observed near the heated wall. Consecutively, a great part of the space offered in the cavity is at a uniform cold temperature. A progressive increase of the emissivity up to 0.85 (highly emissive walls), as shown in Fig. 3b, leads to a complete disappearance of the closed cell located in the right part of the cavity in favor of the open streamlines of the forced flow. This results from the aspiration of the injected flow under the effect of the buoyancy force which develops at the level of the vertical right wall heated under the effect of radiation. The cold zone space is seen to be reduced. This behavior is attributed to the heating of the adiabatic upper and right walls which present increasingly significant thermal gradients as the emissivity increases.

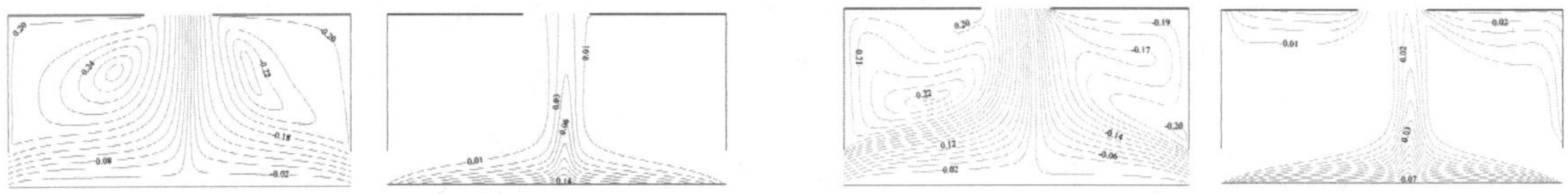

a) (Ψ_{min} = -0.218, Ψ_{max} = 0.250) **b)** (Ψ_{min} = -0.201, Ψ_{max} = 0.224)

Figure 3. Streamlines and isotherms obtained, in the injection mode, for Re = 300 and various values of ε: a) ε = 0 and b) ε = 0.85.

In the case of the suction mode, streamlines and isotherms, illustrating the radiation effect on the flow structure and temperature distribution within the cavity, are presented in Figs. 4a-4b, for Re = 300 and various values of ε. Hence, Fig. 4a, obtained for ε = 0, shows that the flow descends vertically from the upper horizontal inlet and then leaves the cavity horizontally through the vertical two outlets. The structure is characterized by the formation, under the shear effect, of two big closed cells surmounting the open streamlines. These cells are straight and large in size and intensity compared to those presented in the injection mode (see Fig. 3a). The corresponding isotherms illustrate a very limited thermal boundary layer. Consequently, the heat released by the bottom hot wall is quickly and directly transferred to the exit through the thermal boundary layer without ascending under the buoyancy forces effect. It results from this behavior that the suction mode is thermally strong. By increasing progressively ε up to 0.85 (Fig. 4b), the left closed cell is reduced in size and intensity in favor of a small convective cell generated at the upper left corner of the cavity and whose the formation is due essentially to the growing natural convection effect involved by the heating of the upper wall under the radiation effect. In the other hand, the cold zone is relatively reduced.

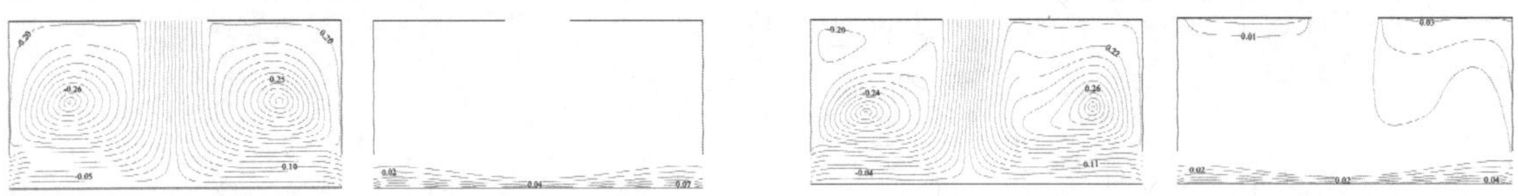

a) (Ψ_{min} = -0.259, Ψ_{max} = 0.263) **b)** (Ψ_{min} = -0.252, Ψ_{max} = 0.270)

Figure 4. Streamlines and isotherms obtained, in the suction mode, for Re = 300 and various values of ε: a) ε = 0 and b) ε = 0.85.

Variations versus Re of the total Nusselt number are presented in Fig. 5. The results show increasing tendencies of Nu_H with Re and ε for the two ventilation types. Quantitatively, in the injection case, for ε = 0.85, improvements in terms of total heat transfer due to surface radiation is about 36 % and 29 % for Re = 300 and 5000 respectively in comparison with the reference case corresponding to ε = 0. Moreover, with respect to the injection mode, the enhancement of heat transfer achieved by the suction mode is very important. For instance, for ε = 0.85 and Re = 5000, passing from the injection mode to the suction one, Nu_H increases from 35.9 to 66.4 which corresponds to an enhancement of the heat transfer by about 84.9 %.

The contribution of radiation to the total heat transfer through the heated wall of the cavity is quantified by presenting in Fig. 6 the evolution of the ratio $Nu_H(rd)/Nu_H$ with Re for various value of ε in both injection and suction cases. It is seen from the figure that, for a given ε, the contribution of radiation decreases, in generally, by increasing Re. However, despite this decrease, the contribution of radiation remains important, especially for moderate and high values of ε. More precisely, the maximum contribution of the radiative component reached in the injection / (the suction) mode for ε = 0.85 and Re = 300 is about 36.88 % / (34.36 %) respectively. Finally, it should be noted that the injection mode is more favorable to the contribution of radiation in comparison with the suction mode.

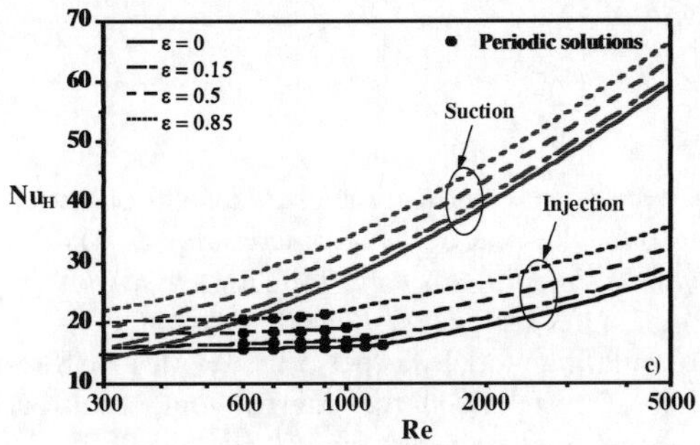

Figure 5. Variations, with Re, of the average total Nusselt number on the heated wall for various values of ε in both injection and suction modes

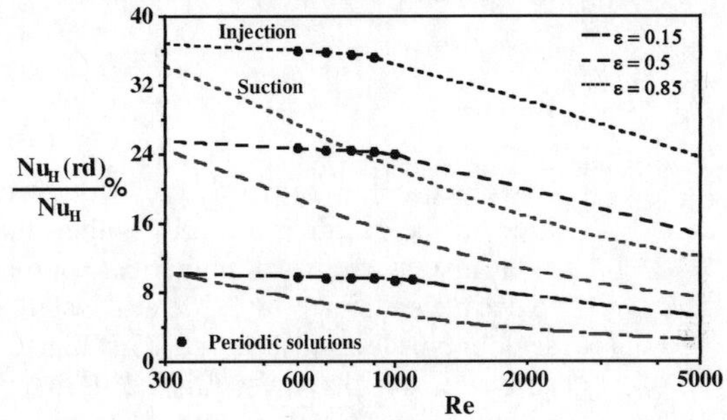

Figure 6. Contribution of radiation evaluated at the heated wall, as a function of Re, to the overall heat transfer for various values of ε in both injection and suction modes.

6. Concluding remarks

In the present work a numerical study was carried out to examine laminar mixed convection coupled to thermal radiation in a vented cavity totally heated from below. The study is conducted by considering two modes of imposed external flows (injection and suction). The obtained results reveal that the surface radiation alters widely the flow and temperature patterns, contributes to the temperature homogenization within the cavity and enhances the overall heat transfer. Also, it is found that the injection mode favors the contribution of radiative component but the suction mode is more efficient than the injection mode by leading to more total heat transfer across the cavity even with or without radiation.

References

[1] Bouafia M, Hamimid S and Guellal M 2015 *Int. J. Ther. Sci.* **96** 236-247
[2] Ridouane E H, Hasnaoui M and Campo A 2006 *Heat Mass Transfer* **42** 214-225
[3] Ramesh N and Merzkirch W 2001 *Int. J. Heat and Fluid Flow* **22** 180-187
[4] Singh D K and Singh S N 2016 *Int. J. Ther. Sci.* **107** 111-120
[5] Raji A and Hasnaoui M 2001 *Engineering Computations* **18** 922-949
[6] Bahlaoui A, Raji A, Hasnaoui M, Ouardi C, Naïmi M and Makayssi T 2011 *J. App. Fluid Mechanics* **4** 89-96
[7] Hottel H C and Saroffim A F 1967 *Radiative heat transfer* (McGraw-Hill, New York)
[8] Wang H, Xin S and Le Quere P 2006 *Comptes Rendus Mecanique* **334** 48-57.

IOP Conf. Series: Materials Science and Engineering **353** (2018) 012002 doi:10.1088/1757-899X/353/1/012002

Comparative experimental and numerical studies of usual insulation materials and PCMs in buildings at Casablanca

Amina MOURID[1], Mustapha EL ALAMI[1]

LPMMAT, Département de Physique, Faculté des Sciences Aïn Chock, Université Hassan II de Casablanca, Morocco

mouridaminamail@gmail.com

Abstract. In this paper, we present a comparative thermal study of the usual insulation materials used in the building as well as the innovate one like phase change materials (PCMs). Both experimental study and numerical approach were applied in this work for summer season. In the experimental study the PCM was installed on the outer surface on the ceiling of one of two full-scale rooms located at FSAC, Casablanca. A simulation model was performed with TRNSYS'17 software. We have established as a criterion of comparison the internal temperatures. An economic study also has been carried out. Based on this latter, that the PCM is most efficient.

Keywords: Insulation materials, Energy efficiency, full-scale room, experimental study, PCMs, TRNSYS'17

1. Introduction

According to the international energy agency, the global energy consumption related to the residential sector in Morocco has almost doubled (47%) between 2004 and 2011[1]. During the summer season, a large part of this energy consumption is related to space cooling [2]. Generally, in residential buildings, the surface that has the largest exposure to the outdoor environment (solar radiation) and, consequently, most subjected to the climate change is represented by the roof [3]. In fact, lightweight, cannot provide the necessary mass to buffer thermal gain. So far, the addition of new insulation layers appears a promising solution to improve the thermal behavior of roof components [4].
Over the last few decades, several studies and models are interested in developing new technologies to improve the thermal behavior of roofs taking into account the limitation of thickness and weight of insulation materials and the height reduction of the internal space. New technology has been developed like the supper insulating [5,6], radiant barriers [7,8], ventilated air cavity [9, 10], green roof [11] and the cool materials [12,13]. Another approach based on the dynamic thermal properties has developed by the implantation of phase change materials [14, 15, 16].

Guechchati et al. [17] conducted a dynamic simulation study using TRNSYS'16 to evaluate the effect of insulation materials on energy consumption of a building located in Oujda (Morocco). Various solutions have been proposed, the authors have concluded that the insulation of the roof coupled with the external wall insulation both with 6 cm of polystyrene reduce the heating load by 16.4% and 19.8% respectively without and with double glazing. yrne et al. [18] investigated the in situ thermal behaviour of a case study building located in Ireland. Both The ceiling and the external wall has been thermally insulated. The authors have compared data before and after the fixing of insulation. The authors concluded that predicted values of heat losses using standard material properties did not reflect the actual values achieved in situ. Indeed, the predicted thermal resistances based on standard materials properties, overestimated the reduction in heat losses through the ceiling by nearly 50% and through the wall by nearly 20%. Furthermore, a thickness of 10 cm of polystyrene led to heat losses reduction of 56% and

35% through the walls and the roof respectively.

Ibrahim et al. [5] examinated the effectiveness wall structure and insulation layer positions for different construction materials for French Mediterranean climate. Both experiment and numerical model was carried out, the best performance is achieved when placing the insulation at the interior wall surface. In general, the aerogel-based insulating coating is most performed compared to the other insulating materials. Elarga et al. [19] investigated the thermal performances of PCM integrated in a residential attic in Torino (Italy). The PCM was applied to a roof a with three different solutions during summer season. The authors conducted that PCM-enhanced components are a promising solution to improve the thermal performance of the roof. The experimental results showed a reduction of the ongoing heat peak load between 13% and 59% depending on the PCM thermal characteristics. Mourid et al. [16] studied the effectiveness of PCM to improve the thermal comfort during summer period, the PCM was installed on the roof, to parameters has been investigate (thickness and the emplacement of PCM). The results indicate that the integration of the PCM on the roof, led a reduction of daily temperature swings (up to 2°C) and heat flux (more than 88%). The most benefits configuration was the emplacement of the PCM on the outer face of the roof with the thickness of 10.52 cm.

This paper reports the development of a lightweight with an integrated of different types of insulation layer aimed at buffering external gains during summer. Both experimental and numerical study are detailed and the assessment of thermal performance and financial study are discussed.

2. Experiment

2.1. Description of the cells

Building rooms were constructed at Faculty of sciences Ain chock Casablanca as detailed earlier in [16]. The cells have an internal dimension of 2.8 m x 2.8m x 2.8 m. The northern wall is equipped with a door of dimension of 1 m×2 m and window 1 m× 1 m. The ratio of the window to wall area in the northern sides is 17%. The thickness of the window glass is 5 mm while the details of building envelopes along with the thermos-physical properties of the building materials are given in Tables 1 respectively.

Table 1. Detail construction of the building and the thermal properties of Building materials Economics and environmental cost of insulation

Building elements	Construction (outside-inside)	U-Heat transfert coefficient ($W/m^2.K$)
Roof	Cement mortar Bitumen Cement mortar Heavy concrete Cement mortar	2.722
Vertical Wall	Cement mortar Yellow brick Air layer Yellow brick Cement mortar	0.531
Floor	Cement mortar Clay Polystyrene Heavy concrete Limestone	0.694

2.2. Climate in Casablanca city
Casablanca is located in the central-western part of Morocco at latitude 33°35'N and longitude 7°36'O. It has a warm and temperate climate (Mediterranean). it's classified as Cwb by the köpper-Geiger system. July and August are the most warmer months of the year with temperature ranging from 24.23°C to 32.44°C.

2.3 Instrumentation
Thermocouples, K-type (2/10mm) were used, they have been verified and calibrated. The accuracy from the thermocouples is about ±0.20 °C.
The meteorological data was measured using vantage Pro 2 weather station. It's installed on the roof of one of the cells. this station allows us to measure the outside temperature (±0.50°C), the humidity, the speed and wind direction (±1.00m/s) and using pyranometer, we measured the global solar radiation (±5%).All the instruments are connected to a data logger, which is in turn connected to a computer.

2.4 Measured weather data
The experimental results were used to study the thermal performance of the roof and to validate the numerical model. Two days of experimental results (from 28th to 29th June 2015) representative of the summer conditions are reported in Figures 1and 2.
In the monitored period, the outdoor temperature was between ~26.24°C (minimum night temperature) and ~31.15°C (maximum daily temperature). During the studied days, the incident global solar radiation reached a peak of 871 W.m^{-2} at 1:00 pm with average value of 466 W.m^{-2}.
The wind speed is characterized by wind velocity varying within 1.3 m/s <v <11.6 m/s, Figure 2. The minimum value was recorded during the night.

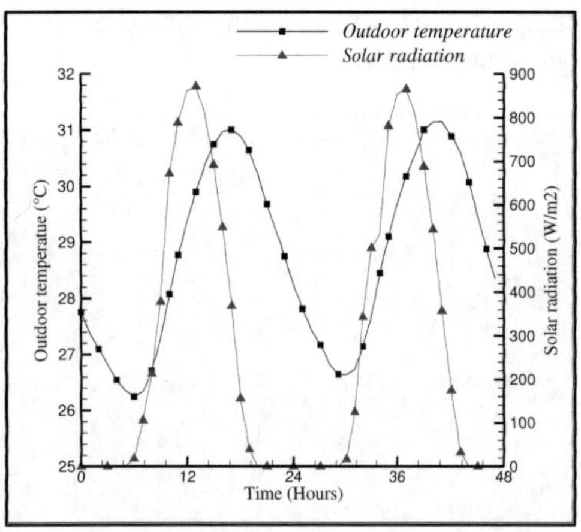

Figure 1. Outdoor temperature and global solar radiation

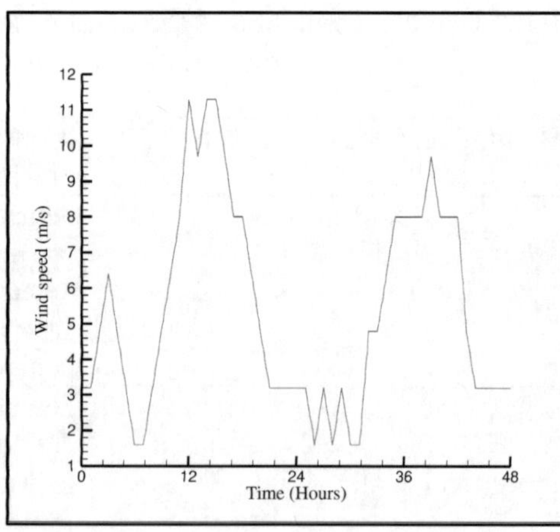

Figure 2. Wind speed variation

2.5 Studied Configuration
In this study different type of thermal insulation were studied. Six different configurations are investigated: The first and the second cases are realized experimentally using the reference cavity with standard walls (Case 1) and the other cavity with PCM on the external face of roof (Case 2). The other

configurations are performed numerically for different kinds of insulation, using TRNSYS'17. These investigations are summarized as follow:
Case 1: reference cavity)
Case 2: PCM panels
Case 3: Rockwall
Case 4: Cork
Case 5: Extruded polystyrene (XPS)
Case 6: Aerogel
Case 7: Vacuum insulation panels
The extruded polystyrene foam, cork, rock with thermal conductivity of 39 mW/(m.K), 40 mW/(m. K) and 41 mW/(m. K) was used as thermal insulation material.
The aerogel used in this study, is called Spaceloft® and developed by Aspen Aerogels [20]. Spaceloft® is a flexible aerogel blanket has a thermal conductivity of 13.1 mW/(m K) at 273 K, 2–2.5 times lower than traditional thermal insulation materials. The Vacuum insulated panels (VIPs) used is fumed silica, it has thermal conductivities of 4mW/(mK) it's called Vacupor® NT-B2 [21].
The phase change material (PCM) used is a mixture of ethylene based polymer (40%) and paraffin wax PCM (60%). The melting temperature of PCM is in the range of 20 °C $\leq T_{melt} \leq$ 35 °C and a heat latent enthalpy h_{PCM} of 72.4 kJ/kg [22]. Its thermal conductivity is 0.18 $W.m^{-1}.K^{-1}$ in solid phase and 0.22 $W.m^{-1}.K^{-1}$ in liquid phase.

3. Simulation details

In order to study the effect of PCM and the insulations Materials on thermal performance and energy consumption for a room building located in Casablanca, the TRNSYS'17 software through transient multi-zone modeling was employed with a step of one hour. The validation of the simulation model is based on the experimental data of two months of measures during summer season. In order to validate this model, meteorological data measured on the site were used. The air temperatures of the cavity obtained from the simulations are compared with those measured. Generally, the comparison of the calculated and the measured temperatures show a good agreement, taking into account the thermocouple accuracy and approximate computational errors. The deviations do not exceed 0.84 °C in cell without PCM (figure 3) and 0.8°C in the cell with PCM (figure 4).

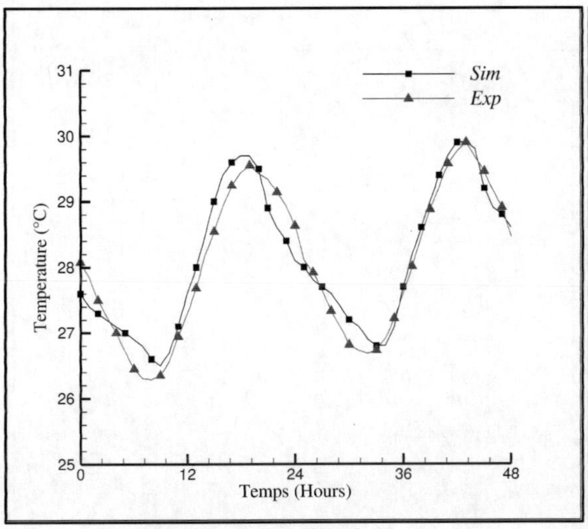

Figure 3. Ref validation

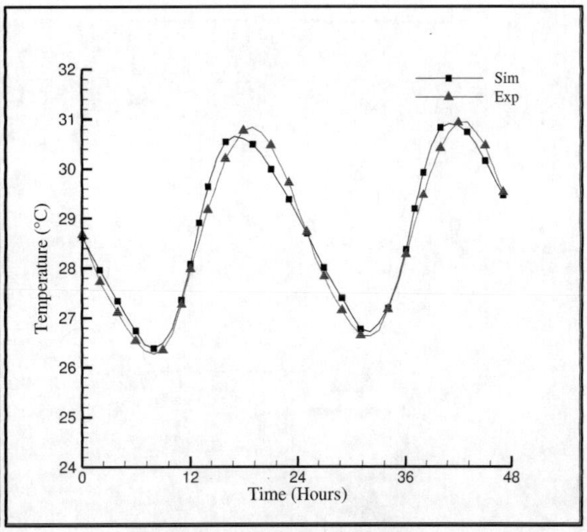

Figure 4. PCM validation

IOP Conf. Series: Materials Science and Engineering **353** (2018) 012002 doi:10.1088/1757-899X/353/1/012002

Figures 5 and 6 present the assembly panel windows of our TRNSYS project for both cells reference and PCM cavity, including all the involved modules (Types) and links.
The cooling loads in the cells are calculated on the basis of the set air temperatures of 21 °C, and the humidity comfort at 60% in summer.

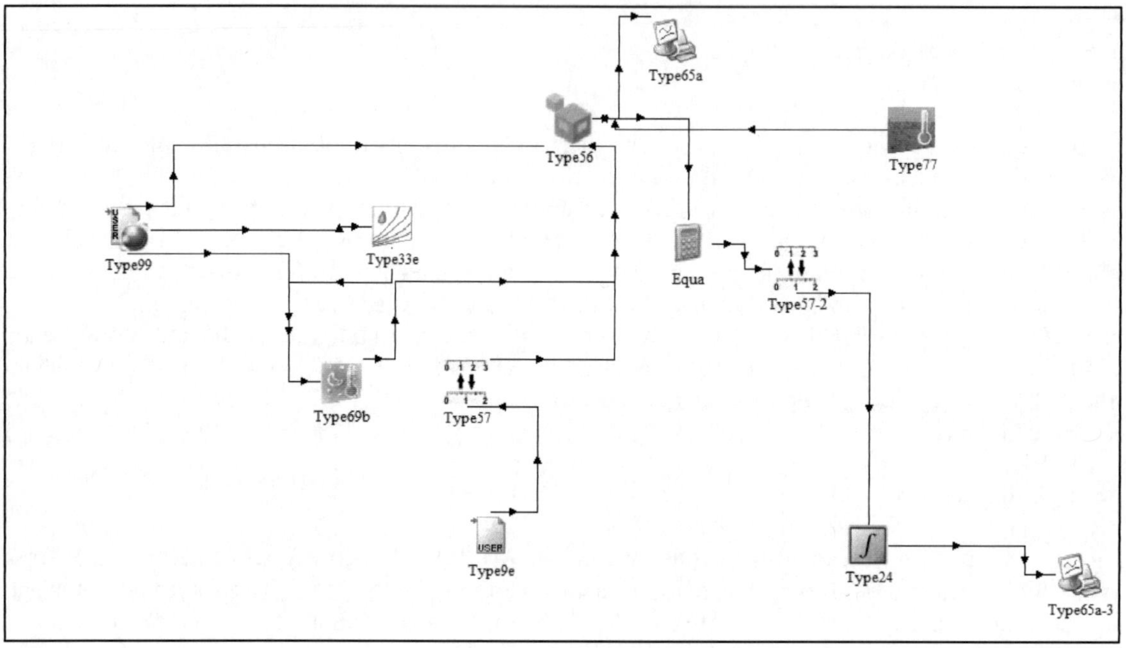

Figure 5. Assembly panel windows of our TRNSYS project for the reference cavity

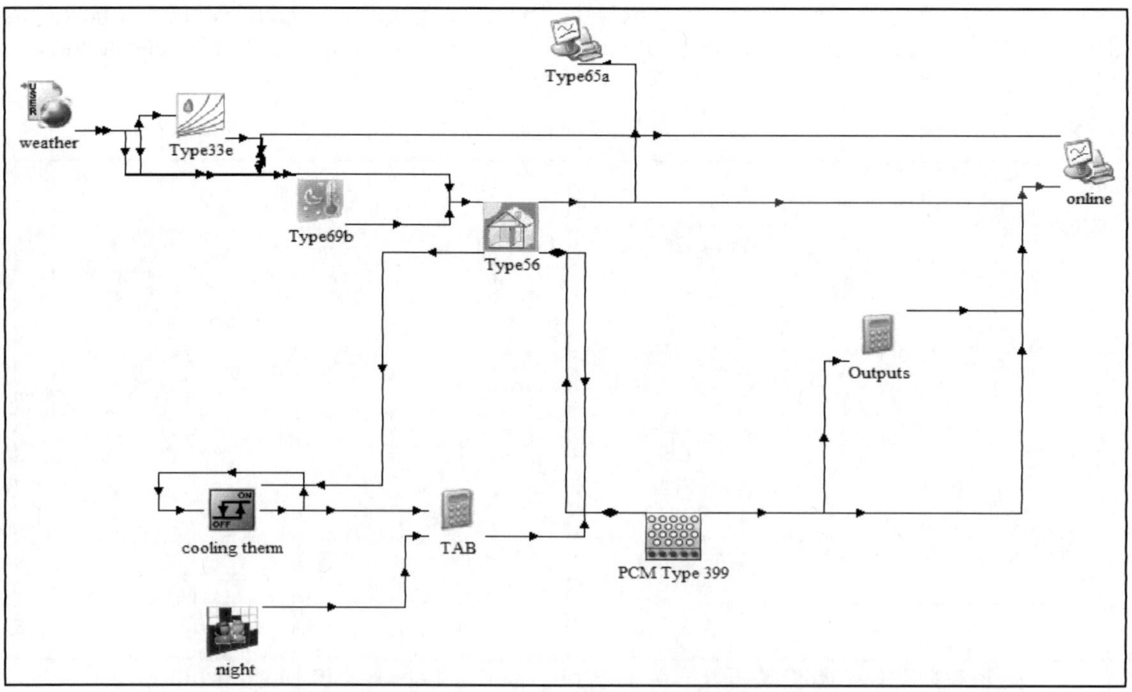

Figure 6. Assembly panel windows of our TRNSYS project for the cell with PCM

IOP Conf. Series: Materials Science and Engineering **353** (2018) 012002 doi:10.1088/1757-899X/353/1/012002

Simulations were performed using the following assumptions:
-Initial air temperature and humidity are taken equal to 20 °C and 50% respectively in studied zone.

- Internal heat generation (Lighting (5W/m^2)).

- The doors and windows are closed all the time (no free cooling).

-Constant infiltration rate of 1 ACH.

- The absorption coefficients of the exterior face of the roof of the reference cavity and the PCM cavity are set to 0.7 and 0.245 [16].

-The wall external emissivity is taken equal to 0.9.

-Convection heat transfer coefficient at the outside surfaces h_{out} is calculated using (Eq 1) which takes into account the wind velocity V [23]:

$$h_{out}=1.44*V + 4.955 \quad (1)$$

-TYPE 77 was used to calculate the ground temperature
-Type 399 models phase change materials (PCM) in wall constructions.

The comparison criteria are the internal temperature and the thickness of the PCM. Indeed, the experimental results are obtained for a PCM thickness of 5.26 mm. The obtained results will be compared with those obtained numerically.

4. Results and discussion

4.1 Experimental results
In this section, the thermal performance of an office building (with and without PCM) located in Casablanca, Morocco is evaluated for a typical two days climate of July 2015. It should be noted that in the case 2 (with PCM application), PCM was incorporated on the outer surface of the roof. The interior temperature profiles for the cavities (with and without PCM) are presented for the considered period in Figure 7. It shows that ceiling surface temperature was reduced by an average of 1.15 °C due to the passive application of PCM with a maximum reduction of 1.28 °C. It can be explained by the increase of the thermal inertia of the roof through the storage and release of energy. Thereby, the thermal behavior of PCM allows reducing the amplitude of temperatures. This figure also shows that PCM temperature varied between 26.28 °C to 29.90 °C during the studied period although the outdoor air temperature doesn’t reach below 21.0°C all considered period. This means that a small part of PCM was solidified at night, which does not allow the PCM layer to work at full capacity during the day.

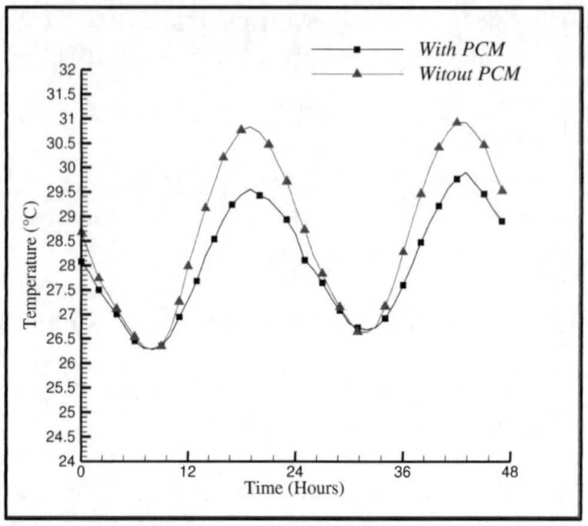

Figure 7. PCM (Case 2)

4.2 Numerical results

Since our experimentation does not allow a complete study with different insulating materials, we have therefore proceeded numerically to complete our comparative study. Indeed, for the numerical configurations, we present the temporal evolutions of the temperatures of these different cases compared to that of Case 2. The curve relating to the insulation for each case is obtained for an optimal thickness, which must be established. This thickness is obtained in the following way: we start the calculations with an exhaustive thickness of insulation that we start to reduce with a step of 2 cm. We compare the curve obtained with that of case 2 for each step. Once we find that the curves start to approach, we reduce the pitch to 1 mm and we continue to reduce the thickness until the two curves are almost identical. In addition, we also verify that we have almost the same energy consumption of the cooling system. Once these two criteria are verified, we consider the thickness as optimal.

In Figures 8, 9, 10, 11 and 12, we present a comparison of the internal temperatures of Case 2 and Cases 3, 4, 5, 6 and 7, respectively. These curves are obtained for the following optimal thicknesses, Table 2:

Table 2. Optimal thicknesses of considered insulating materials

Insulation material	PCM	Rockwall	Cork	XPS	Aerogel	VIP
Thicknesses (mm)	5.26	15	15	12.5	8	6

It should be remembered that the trend for recent constructions is in the direction of light materials as well as thin but heavy envelopes (in terms of thermal mass). For all these reasons, we find that cases 3, 4 and 5 have relatively large optimal thicknesses and should be considered out of competition.

The Other materials of cases 6 and 7 (Aerogel and VIP) have low optimal thicknesses. They can be considered for new buildings. However, the thickness of case 2 (case with PCM) remains the lowest and therefore, we conclude that our phase change material remains the best current insulator by its thermal inertia and its ability to store heat in latent form.

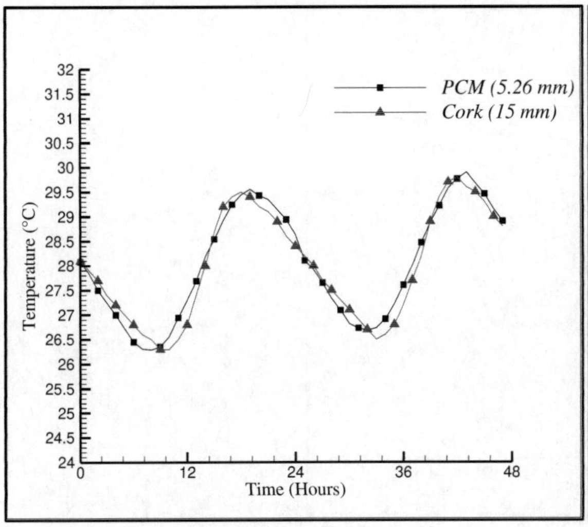

Figure 8. Rockwall (Case 3)

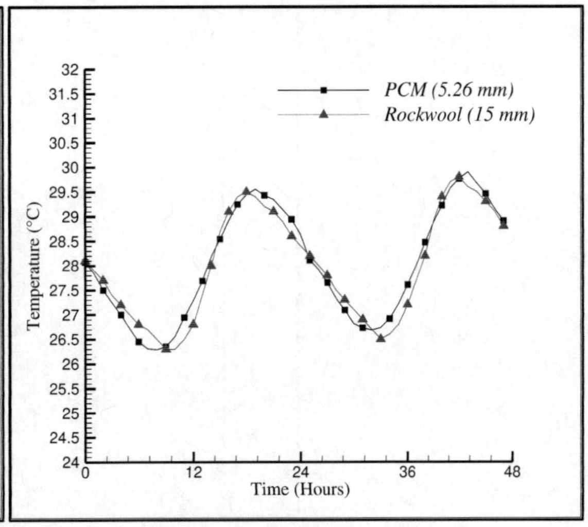

Figure 9. Cork (Case 4)

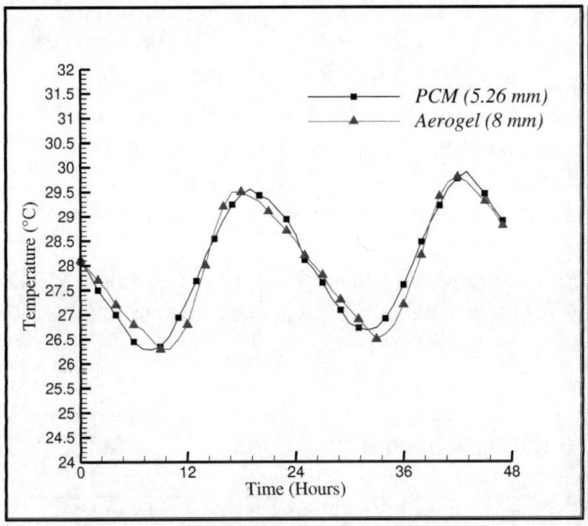

Figure 10. Polystyrene (Case 5)

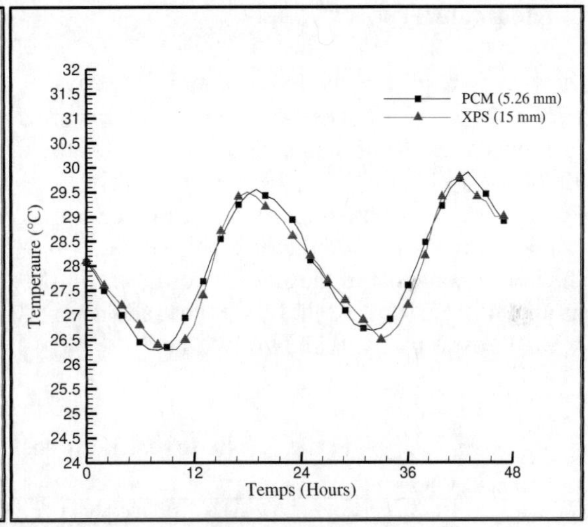

Figure 11. Aerogel (Case 6)

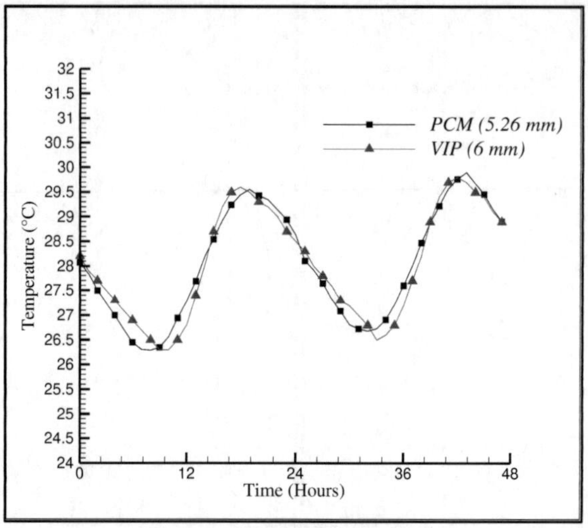

Figure 12. VIP ((Case 7)

5. The financial study

The energy requirements for cooling of the cavities is calculated by TRNSYS'17 software, with a set point of 21°C, as reported above. Simulations are conducted using measured meteorological data of Casablanca. Conventional energy demand for the cavities with or without insulation is about 40878,7095 kJ and 48109, 78 kJ respectively.
Since we set the internal temperature and energy consumption right from the start, so the investment cost will be the decisive factor in this study.
The investment cost of the studied configuration is about 2.000 MAD (around 175 EURO), 4.420 MAD (around 400 EURO), 9.090 MAD (around 822 EURO) for the Case 2 (PCM), Case 6 (Aerogel) and Case 7 (VIP), respectively as shown is table 3.

Table 3. The investment cost of the studied configuration

Insulation types	Insulation material cost (MAD) (m^2)	Cost of making (MAD)	Total cost (MAD)
PCM	331,66	30	2000
Aerogel	763,7	30	4420
VIP	1207,33	30	9090

6. Conclusion

The aim of this paper is to provide experimental and numerical data in order to evaluate a thermal behavior and financial benefit of PCM and insulation materials in the building. Two test cells were used to measure the indoor temperatures of the cells subject to the real conditions of Casablanca. A numerical study was involved using Trnsys'17. In order to validate the numerical model, two steps of validation were presented. The first step consists in comparing with experimental data related to the reference cavity and the second one was with the cavity with PCM.
The results show that the PCM application is the most efficient compared to the supper insulation as Aerogel and VIP. Indeed, it represents the thinnest thickness (5.56 mm) and so the cheapest investment (175 EURO).

IOP Conf. Series: Materials Science and Engineering **353** (2018) 012002 doi:10.1088/1757-899X/353/1/012002

7. References

[1] International Energy Agency (AIE) 2014, OECD/IEA, MAROC 2014

[2] Agence de Développement des Energies Renouvelables et de l'Efficacité Energétique ADEREE, (2012), Règlement Thermique de Construction au Maroc.

[3] U.S. Department of Energy, 2011 Buildings Energy Data Book, 2012, p. 286.

[4] Ballarini I, Corrado V, 2012n Analysis of the building energy balance to investigate the effect of thermal insulation in summer conditions, Energy and Buildings 52 p168–180. `

[5] Ibrahim M, Biwole P H, Wurtz E, Achard P, 2014 A study on the thermal performance of exterior walls covered with a recently patented silica-aerogel-based insulating coating, Building and Environment 81 p 112-122.

[6] Voellinger T, Bassi A, Heitel M, 2014 Facilitating the incorporation of VIP into precast concrete sandwich panels, Energy and Buildings, 85 p 666-671.

[7] Fontaninia AD, Pr'Outa K M, Kosnyb J, Ganapathy subramaniana B, 2016, Exploring future climate trends on the thermal performance of attics: Part 1 – Standard roofs, Energy and Buildings 129 p 32–45.

[8] Escudero C, Martin K, Erkoreka A, Flores L, Sala J M, 2013 Reflective thermal insulation systems in building: A review on radiant barrier and reflective insulation, Energy and Buildings ,59 p 62-72.

[9] Sunwoo L, Park S H, Myong S Y, Kwang W K, 2009, An experimental study on airflow in the cavity of a ventilated roof, Building and Environment, 44, Pages 1431-1439.

[10] Fantucci S, Marinosci C, Serra V, Carbonaro C, 2017, Thermal Performance Assessment of an Opaque Ventilated Façade in the Summer Period: Calibration of a Simulation Model through in-field Measurements, Energy Procedia, 111, p 619-628, https://doi.org/10.1016/j.egypro.2017.03.224.

[11] Alvizuri J, Cataldo J, Smalls-Mantey L A, Montalto F A, 2017, Green roof thermal buffering: Insights derived from fixed and portable monitoring equipment, Energy and Buildings 151, p 455-468.

[12] Xiang B, Xiuping Y, Zhang J, 2017, A novel cool material: ASA (acrylonitrile-styrene-acrylate) matrix composites with solar reflective inorganic particles, Composites Science and Technology, 145, p 149-156.

[13] Ong KS, 2011, Temperature reduction in attic and ceiling via insulation of several passive roof designs, Energy Conversion and Management 52, p 2405-2411. http://dx.doi.org/10.1016/j.enconman.2010.12.044.

[14] Min Hee, C. Jin Chul, P.2016, Development of PCM cool roof system to control urban heat island considering temperate climatic conditions, Energy and Buildings,116, p 341-348.

IOP Conf. Series: Materials Science and Engineering **353** (2018) 012002 doi:10.1088/1757-899X/353/1/012002

[15] Mourid, A, El Alami, M, 2016, Passive Study of Energy Efficiency of a Building with PCM on the Roof during Summer in Casablanca, Journal of Power and Energy Engineering, 4, p 26-37.

[16] Mourid A, Alami M, 2017, Thermal Behavior of a Building Provided with Phase Change Materials on the Roof and Exposed to Solar Radiation, Journal of Solar Energy Engineering 139(6).

[17] Guechchati R, Moussaoui M A, Mezrhab Ahm, Mezrhab Abd, 2010, Simulation de l'effet de l'isolation thermique des bâtiments : Cas du centre psychopédagogique SAFAA à Oujda. Revue des Energies Renouvelables,13, p 223-232.

[18] Byrne A, Byrne G, Davies A, Robinson A J, 2013, Transient and quasi-steady thermal behavior of a building envelope due to retrofitted cavity wall and ceiling insulation, Energy and Buildings 61 p 356-365.

[19] Elarga H, Fantucci S, Serra V, Zecchin R, Benini E, 2017, Experimental and numerical analyses on thermal performance of different typologies of PCMs integrated in the roof space, Energy and Buildings 150, p 546-557

[20] Spaceloft® Safety Data Sheet, Retrieved November 19, 2009, from http://www.aerogel.com/products/pdf/Spaceloft MSDS.pdf.

[21] Vhttp://www.porextherm.com/en/products/vacupor.html

[22] Kuznik ,F, Virgone, J, Johannes, K, 2010, Development and validation of a new TRNSYS type for the simulation of external building walls containing PCM. Energy and Buildings, 42, p 1004–1009

[23] Mourid, A, El Alami, M, 2017, Experimental analysis of the thermal behavior of two cavities kind of living space with and without PCM on envelopes, *International Journal of Modern Embedded System, 5, p 7-12 ISSN: 2320-9005.*

On the application of a new thermal diagnostic model: the passive elements equivalent in term of ventilation inside a room

El Mehdi EL KHATTABI[1], Mohamed MHARZI[1], Saad Raefat[2], Zouhair MEGHARI[3]

[1] Laboratory for Integration of Advanced Systems and Technologies-University Sidi Mohamed ben Abdellah-Faculty of Sciences Dhar El Mehraz, Fez, Morocco
[2] Laboratory of Energy, Materials and Environment, Mohammed V University in Rabat - EST Salé, 11060, 227 Avenue Prince Héritier, Salé, Morocco
[3] Department of Physics, Faculty of Science, Moulay Ismail University, Meknes, Morocco

mehdienerg@gmail.com

Abstract. In this paper, the thermal equivalence of the passive elements of a room in a building located in Fez-Morocco has been studied. The possibility of replacing them with a semi-passive element such as ventilation has been appraised. For this aim a Software in Fortran taking into account the meteorological external conditions along with different parameters of the building envelope has been performed. A new computational approach is adapted to determinate the temperature distribution throughout the building multilayer walls. A novel equation gathering the internal temperature with the external conditions, and the building envelope has been deduced in transient state.

Keywords: Passive elements; Thermal performance; Building envelope; Dynamic state

1. Introduction

The Thermal calculation accuracy. Affects the thermal behavior of any building and the design of heating and cooling loads required to maintain thermal comfort. Assessing the amount of heat lost through external walls is a prerequisite for building energy. However, the method of estimating heat loss in the building creates a significant difference, which leads to disorder of the building design . As such, Several previous works concerning the study of the thermal behavior were realized during the last decade such as the influence of the thermal properties of the material and its thermal characteristics.

Al Sanea et al [1] developed a thermal model script model based on the implicit method in infinite volume, to evaluate the thermo-physical properties of the building. The concept of thermodynamic resistance (R-value) has been developed to calculate the effects of wall orientation, radiative exchange, and thermal energy storage as well as heat resistance. This parameter is also based on several variables such as building type, weather conditions and performance of material properties, that is treated in [2-5]. Other studies such as Tsilingiris PT et al [6] studied the effect of diffusion of heat capacity and thermal resistance on the behavior of a wall under varying initial conditions. The results showed that the use of large walls with a thermal insulation capacity of the room leads to an increase in the thermal time constant; it characterizes the rate of change of the temperature inside the building compared to the external conditions. Sami A. Al-Saneaet al [7] have also theoretically studied the effects of thermal mass on the dynamic heat transfer characteristics of building walls with the same thermal resistance (R-value). K.j.kontelion et al [8] have discussed the effect of construction parameters taken into account the position of the insulation layer inside the wall. Philipps et al [9] assessed the ability to search a new combination of thermal mass model collected, accompanied by a Bayesian statistical analysis to estimate the U- value of thermal mass. Furthermore, Walker et al [10] have also analyzed the thermal performance in a dynamic state of seven internal insulation options, by experimental method using a heat flow sensor.

2. Numerical modelling

The theoretical analysis has been made under transient-state conditions. The aim of this numerical modeling is to combine the different parameters and conditions affecting the internal temperature of building in order to determine accurately its variation in function of time, starting from the different external phenomena, passing by the envelope through the determination of the internal walls temperature and finishing by coupling the entire internal parameters inside building. The numerical model takes into account the impact of the external atmospheric conditions along with the envelope composition coupled with internal building parameters. The isotropic solar model [11-13] was adjusted to evaluate the solar irradiation for each iteration using the same step of time Δt.

Internal temperature determination

In order to evaluate precisely the building internal temperature, the most important heat gains must be taken into account. In this work, supposing that the building isn't occupied, we neglect the occupants and equipment's heat gains and we only consider parameters related to the temperature variation, namely; the total flux, conduction and radiation, through building envelope (roof, walls, window…) and the air exchange heat between the interior of the building and the outside, thus, the next equation is considered:

$$\phi_{sto} = \phi_{env} + \phi_w + \phi_{ac} \qquad (1)$$

With: ϕ_{sto}, ϕ_{env}, ϕ_w, ϕ_{ac} are respectively, the total heat gain inside building, the conductive heat flow through envelope, the conduction and radiation flux through windows and the air exchange rate between the inside and outside:

$$\frac{m_{air}.c_{air}}{\Delta t}.(T_{in}^{k+1} - T_{in}^{k}) = \left(\sum_{i=1}^{n} \frac{S_i}{R_{ci(n+1)}}\right).(T_n^k - T_{in}^k) + F_S.G_w^k + U_w.S_w.(T_{ex}^k - T_{in}^k) + \left(\frac{\rho_{air}.c_{air}.Q_v}{3600}\right).(T_{ex}^k - T_{in}^k) \qquad (2)$$

With: n, S_i, $R_{ci(n+1)}$, T_n^k, F_s, G_w^k, U_w, S_w and Q_v are respectively: the total number of the building external partitions, the surface of the considered partition, the partition inner thermal resistance. the inner layer temperature of the iteration k determined using the discretization model. The solar factor, the direct solar radiation on windows of the iteration k calculated using the solar model, the total thermal transmittance of windows total surface and the air flow rate.

3. Results and discussion

3.1. Description of the studied case:

The thermal study of a room in a modern building design(figure 1)located in Fez has been carried out, taking into account meteorological data and thermo-physical properties of the building materials

[14] (table 1 and table2), to assess in one hand the impact of the orientation of the glazing, the insulation of the room and the glazing rate, on the interior temperature variation in the studied room, and in the other hand to calculate the equivalent air flow rate maintaining the same interior temperature which correspond to each studied passive element. Results are illustrated in Figure 2.

Table 1. Layers and thermo-physical properties of exterior walls

	Thickness (cm)	Thermal capacity (kJ/kg K)	Thermal conductivity (W/m K)	Mass density(Kg/m³)
polystyrene	2-12	1.380	0.00	˙0035
Cement mortar	1.5	1.000	1.15	1700
Brick	20	0.878	1.15	1800

Table 2. The layers and thermal properties of the roof

	Thickness (cm)	Thermal capacity (kJ/kg K)	Thermal conductivity (W/m K)	Mass density (Kg/m³)
Floor tile	02	0.70	1.750	2300
Mortar	10	1.00	1.150	1700
Concrete	04	0.65	1.750	2100
hourdis	16	1.00	1.230	1300
Plaster	02	1.00	0.335	1500

Figure 1. The building where the studied room is located a- South/East side b- North /West side.

In the first case (figure 2-a and figure 2-b) the equivalence of the insulation thickness impact on the thermal temperature in term of air rate flow has been evaluated for each thickness, for example, maintaining the same external meteorological conditions, in the case of 20cm of insulation thickness (figure 2-a), no insulation is considered in figure 2-b and we calculate the amount of air rate flow able to ensure the same daily internal temperature inside the room. The same concept is applied in order to replace the effect of the glazing rate and the impact of the room orientation with a controllable semi-passive element which represented in this study by ventilation.

IOP Conf. Series: Materials Science and Engineering **353** (2018) 012003 doi:10.1088/1757-899X/353/1/012003

As shown in Figure 2-a and Figure 2-b at an insulation thickness of 12 cm of the expanded polystyrene, the temperature of the interior air of the room reaches the same state of equilibrium whatever the external fluctuations.

In Figure 2-c and Figure 2-d the window area changes proportionally with the global solar irradiance transmitted to the room, which explains the indoor temperature increase for each glazing rate.

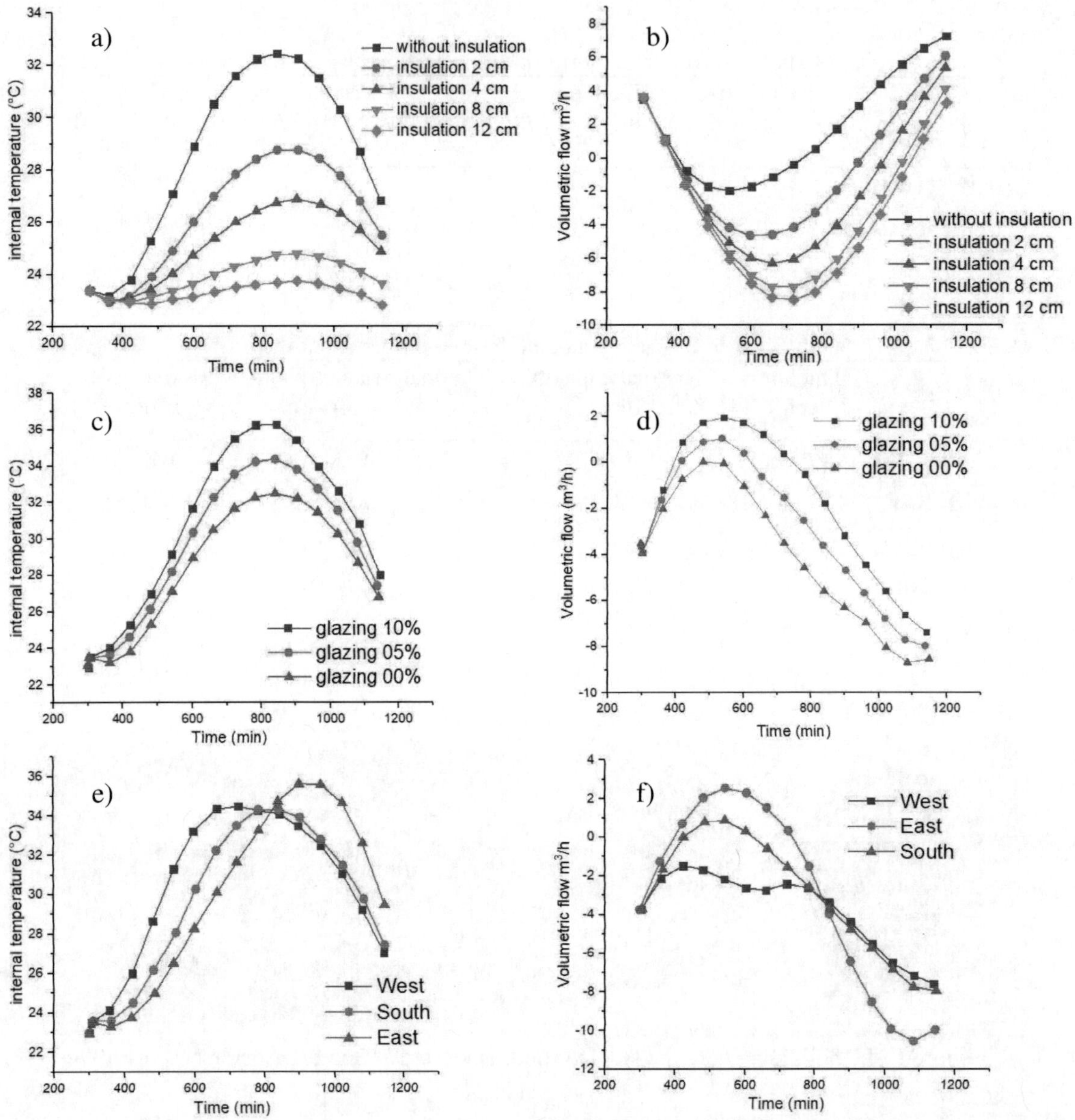

Figure 2. Impact of the passive elements on the daily interior temperature (1 June, Fes) of a room: a-Insulation effect b- Equivalence of air rate flow for each insulation thickness. c-Glazing rate effect d. equivalence of air rate flow for each glazing rate. e-Orientation effect f- equivalence of air rate flow for the different orientations.

In Figure 2-e and Figure 2-f the difference is significant between East and West, in the East direction the morning sun causes a rise in temperature, creating a state of overheating that is maintained throughout the day. However, in the West, the solar radiation arrives only at the end of the afternoon, indeed, the south remains an intermediate orientation between the East and the West.

4. Conclusion

In this paper, we have presented several thermal simulations of a room in a building located in Fez, using a new numerical modeling. We have proposed an equivalence technique of some passive elements by a semi passive element which consisted of ventilation. The evolution of the ventilated air flow rate in one day was calculated in order to maintain the same interior temperature when we use passive elements (orientation, wall insulation and glazing rate). This method has the following advantages:

- ✓ Complete control of the extracted air flow in the case of a forced ventilation.
- ✓ Reduction of the risk of moisture encountered in walls and roofs.
- ✓ Highly controllable system independently of the external weather conditions.
- ✓ Very good air distribution insuring a thermal comfort inside buildings.
- ✓ Maintenance of the interior comfort even with the aging of the building.

As perspective, this study must be extended to one complete year instead of one day and in many different areas, this will prove that the proposed promoting concept could ensure a continuous thermal comfort inside all building types (new and old dwellings) regardless of the passive elements that remain difficult to control especially after long periods.

References

[1] Al-Sanea SA. Evaluation of heat transfer characteristics of building wall elements. J King Saud Univ, Eng Sci 2000;12(2):285–313.
[2] BRE Report, Field Investigations of the Thermal Performance of Construction Elements as Built 36/8/79 cc1637 Final Report, 2000.
[3] Kontoleon KJ, Bikas DK. The influence of the zone's indoor temperature settings on the cooling/heating loads for fixed and controlled ventilation. Building and Environment 2006;41(2):75-86
[4] W.J. Cheng, H.L. Lee, A simulation-optimization approach for energy efficiency Of buildings,Int.J.Energy Manage.49(2006)2117–2121.
[5] E. M. El Khattabi and M. Mharzi, "Effect of locations and thicknesses for the different material constituting a building wall," Energy Procedia, vol. 139, pp. 328–333, Dec. 2017.
[6] Tsilingiris PT. Parametric space distribution effects of wall heat capacity and thermal resistance on the dynamic thermal behavior of walls and structures, Energy Build 2006;38:1200–11.
[7] Al-Sanea SA, Zedan MF. Effect of thermal mass on performance of insulated building walls and the concept of energy savings potential,Applied Energy 89 (2012) 430–442
[8] K.J. Kontoleon, E.A. Eumorfopoulou. The effect of the orientation and proportion of a plant-covered wall layer on the thermal performance of a building zone, Building and Environment 45 (2010) 1287–1303.
[9] Phillip Biddulph, Virginia Gori, Clifford A. Elwell. Inferring the thermal resistance and effective thermal mass of a wall using frequent temperature and heat flux measurements,Energy and Buildings 78 (2014) 10–16.
[10] Rosanne Walker, Sara Pavía. Thermal performance of a selection of insulation materials suitable for historic buildings, Building and Environment 94 (2015) 155e165
[11] Duffie, J. A., Beckman, W. A., Solar Energy Thermal Processes, 2nd ed., John Wiley Interscience, New York,USA, 1974.
[12] Kasten, F., The Linke Turbidity Factor Based on Improved Values of the Integral Rayleigh Optical Thickness,Solar Energy, 56 (1996), 3, pp. 239-244

IOP Conf. Series: Materials Science and Engineering **353** (2018) 012003 doi:10.1088/1757-899X/353/1/012003

[13] Jannot, Y., Solar Thermal, Courses and Exercises, Chapter: Solar Energy, Chapter: bibliography: Radiation Geo- metrical Form Factors, October 2003
[14] Document, "TRNSYS, Version 16 User Manual".

Solar air heating system: design and dynamic simulation

M. BOUOUD [1], O. HACHCHADI[1], K. JANUSEVICIUS[2], V. MARTINAITIS[2], A. MECHAQRANE[1]

[1]Sidi Mohamed Ben Abdellah University, Electrical Engineering Department, Faculty of Sciences and Technology, Renewable Energies and Intelligent Systems Laboratory BP 2202 Fez, Morocco
[2]Laboratory of Building Energy and Microclimate Systems, Faculty of Environmental Engineering, Vilnius Gediminas Technical University, Saulėtekio al.11, Vilnius, Lithuania.

mahmoud.bououd@usmba.ac.ma

Abstract. The building sector is one of the big energy consumers in Morocco, accounting for about 23% of the country's total energy consumption. Regarding the population growth, the modern lifestyle requiring more comfort and the increase of the use rate of electronic devices, the energy consumption will continue to increase in the future. In this context, the introduction of renewable energy systems, along with energy efficiency, is becoming a key factor in reducing the energy bill of buildings. This study focuses on the design and dynamic simulation of an air heating system for the mean categories of the tertiary sector where the area exceeds 750 m3. Heating system has been designed via a dynamic simulation environment (TRNSYS) to estimate the produced temperature and airflow rate by one system consisting of three essential components: vacuum tube solar collector, storage tank and water-to-air finned heat exchanger. The performances estimation of this system allows us to evaluate its capacity to meet the heating requirements in Ifrane city based on the prescriptive approach according to the Moroccan Thermal Regulation. The simulation results show that in order to maintain a comfort temperature of 20°C in a building of 750m3, the places requires a thermal powers of approximately 21 kW, 29 kW and 32 kW, respectively, for hotels, hospitals, administrative and public-school. The heat generation is ensured by a solar collector areas of 5 m², 7 m² and 10 m², respectively, for hotels, hospitals, administrative and public-school spaces, a storage tank of 2 m3 and a finned heat exchanger with 24 tubes. The finned tube bundles have been modelled and integrated into the system design via a Matlab code. The heating temperature is adjusted via two controllers to ensure a constant air temperature of 20°C during the heating periods.
Keywords: Solar energy, heating system, tertiary buildings, heat exchanger, vacuum tube collectors.

1. Introduction

At the global level, the building sector accounts for around 28% of final energy consumption and contributes about a third of CO_2 emissions. In Morocco, this sector is the largest consumer of primary energy, 23% of the total consumed energy, which represents 17% for residential and 7% for the tertiary sector. This energy consumption is expected to increase in the coming years due to:

- The significant evolution of the building stock through the setting up of different programs: Azur plan for hotels, emergency program for national education, program of 150000 housing units per year and hospital rehabilitation program.

IOP Conf. Series: Materials Science and Engineering **353** (2018) 012004 doi:10.1088/1757-899X/353/1/012004

- The modern lifestyle requirements and the low price of equipment (heating systems, air conditioning, water heating, refrigeration ...) [1].
Significant research and developments have taken place in the last 20 years, in terms of the energy efficiency in buildings, whether for thermal isolation materials, building architectures, electrical equipment and also on energy conversion systems.
In terms of thermal insulation, significant evolutions have been observed, this is the case of low-emissivity double glazing, led to a two-thirds reduction in thermal losses. Thus, the appearance of selective layer glazing, makes it possible to limit solar gains and avoid overheating [2].
The technologies and systems related to the ventilation of buildings have been the subject of significant advances: advent of DC ventilators and double-flow ventilation, allows a considerable increase in the consumption of ventilation systems.
In terms of lighting, the technological evolution is not left out, new lamps have appeared which have a greater luminous efficiency. These include compact fluorescent lamps, which at equivalent luminous flux, divide the installed power by three, as well as LEDs (Light Emitting Diodes), mainly used for signal lighting, which help to reduce the consumption of lighting installations. Although the use of electronic ballasts instead of magnetic ballasts leads to a reduction in the consumption of the installations, thanks to the fact that the control systems are taken into consideration. [3]
In the heating sector, progress is also sensitive. In fact, the efficiency of the heat production units has increased significantly with the appearance of low and very low temperature boilers, condensing boilers and solar water heaters. Improvements in burner technology have resulted in an improved efficiency and reduced emissions of CO_2. Regulation system has also evolved thanks to the introduction of electronic modules, which have made it possible to remote control or diagnosis. Point-to-point control techniques (eg thermostatic valves) have become widespread and have been combined with centralized controls.
Reducing energy consumption in buildings is therefore a major economic and ecological issue. It is in this context that the considered solar air heating system, comes to improve energy efficiency in the building sector especially in the tertiary sector trough the integration of solar energy.
In this article, we will show the calculation methods used, the design and the simulation of the proposed system. The main objective of this system is to maintain a temperature of 20°C inside the building, by adjusting the injected air flow and using solar collectors.

2. Methodology

Based on the seasonal variations of solar irradiation and consequently ambient temperature the required thermal energy in a building has to be estimated in order to optimize the air heating system capacity and consequently ensure the comfort temperature of 20°C.
According to the Moroccan Thermal Regulation in Building, there are two methods to define the required thermal power:

2.1. Performance approach

The performance approach consists of defining the minimum technical specifications in terms of thermal performance of the building, these are evaluated through the annual energy needs of the building related to thermal comfort. In our case, the heating requirements, according to the Moroccan thermal regulation, are about 50 kWh/m²/year for the good insulation conditions and 120 kWh/m²/year for poor conditions in the climatic zone for Ifrane city.
The required power ensured by the proposed solar heating system can be estimated as follows:

$$\boldsymbol{P = H_L \times \Delta T} \qquad (1)$$

With:
P: Thermal power of the system [W]
$\mathbf{H_L}$: Heat losses [W/°C]

IOP Conf. Series: Materials Science and Engineering **353** (2018) 012004 doi:10.1088/1757-899X/353/1/012004

ΔT: Difference of the inside temperature and the outside temperature (**T_{int}** – **T_{ext}**) [°C]

Heat losses are calculated using this formula:

$$H_L = \frac{Q_{ann}}{UDD} \quad (2)$$

With:
Q_{ann}: Annual Heat Requirements [kWh/yr]
UDD: Unified Degree Days [°CJ]

An empirical correlation has been established from the measurements of 146 weather stations to estimate the unified degree days [4]:

$$UDD = -100{,}51 \times T_{ext} + 1704 \quad (3)$$

With: **T_{ext}**: Outside temperature [°C]

2.2. Prescriptive approach

The prescriptive approach consists of setting the acceptable technical specifications limit according to the walls thermal characteristics of the building envelope. These characteristics are represented by the volume thermal transmittance for tertiary sector based on the Moroccan thermal regulation [5]. Therefore, the required thermal power is calculated as follows:

$$P = G \times V \times \Delta T \quad (4)$$

With:

P: Thermal power of the system [W]
G: Volume thermal transmittance [W/(m^3×C°)]
V: Volume of the building [m^3]
ΔT: Difference of the inside temperature and the outside temperature (T_{int} – T_{ext}) [°C]

The prescriptive approach is considered in the coming simulations because of its precise estimation of the energy needs for a given space and thermal transmittance based on the Moroccan Thermal Regulation.

3. Control and design parameters

The solar heating system design has been conducted in a dynamic simulation environment (TRNSYS'17) to estimate the generated air temperature and airflow rate by one system consisting of three main components: flat plate solar collector, storage tank and water-to-air finned heat exchanger as it is show in figure 1. The heating system has been designed to ensure a comfort temperature of 20°C and to compensate the thermal heat losses from September 22th to March 19th from a volume of 750m^3 in different categories of tertiary sector based on the minimal thermal transmittance recommended by the Moroccan regulation namely: administrative and public school buildings (G=1.78 W/$m^{3°}$C), hospitals (G=1.17 W/$m^{3°}$C) and hotels (G=1.47 W/$m^{3°}$C). Different control functions have been considered according to the daily required heating hours for each category. Hotels and hospitals need to be heated all the time, however, the heating period is from 8 am to 6 pm in administrative and public school buildings.

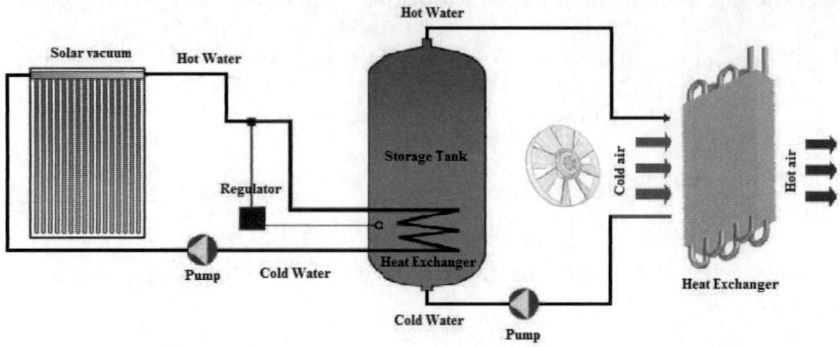

Figure 1: Solar heating system

The heating system sizing has shown that in order to satisfy the required air flow rates and temperature, the flat plate solar collector has to be sized based on the building category and the maximal water flow rate in the solar loop should be 1kg/s, accumulated in a water tank of $2m^3$. Furthermore, the hot water circulates in the heat exchanger loop with a maximal flow rate of 0.5kg/s where the thermo-flow characteristics have been estimated based on the last established models in the literature [6]. The hourly solar irradiations in Ifrane city have been obtained from Meteonorm software and considered in the present dynamic simulations. In order to control the outlet air temperature, two hysteresis controllers have been used in the solar heating system. The first controller was configured in the solar loop based on an upper dead band of 10°C and lower dead band of 2°C while the second one was placed in the heat exchanger loop based on the high limit cut-out temperature with the aim to stabilize the outlet air temperature at 20°C and stop the heat exchanger blower when the ambient temperature exceeds 20°C. Nevertheless, regarding the intermittent behaviour of solar radiation an additional heater was necessary to compensate the possible energy deficit.

4. Results and discussions

4.1. Heat flow rate

The heat losses from a space of 750 m^3 has been calculated for different building categories based on the thermal transmittance for each one and compared with the total heat flow rate provided by the solar heating system.

Figure 2 shows a comparison between the total heat flow rate, provided by a solar collector of $5m^2$, and the heat losses from a Hotel space. It can be seen that the considered system is able to produce more than 26 kW that is enough to cover the maximal required heat which is around 21 kW. In the worst case, the additional heater contributes with not more than 2.4 watts which represents 0.01% of the maximal heat flow rate.
Figure 3 illustrates that a hospital space of 750 m^3 requires a solar collector area of 7 m^2 providing a maximal thermal flow rate of around 29 kW in order to cover the maximal heat losses of around 26 kW. Only 0.005% of the maximal heat flow rate is covered by the additional heater generates.
Figure 4 shows that the total thermal flow rate generated by the solar heating system satisfy the required heat in the administrative and public school spaces using a solar collector area of 10 m^2. In this estimation, an air flow rate of 1.3 kg/s has been considered to compensate the maximal heat losses from the heated space which is estimated by around 32 kW.
As it can be observed in the previous figures, the solar heating system generates higher power in some periods which can be adjusted by controlling the air flow rate according to the required thermal flow rate. Generally, the heat generation is stopped in two cases, the first one is when the ambient temperature is more than 20°C as it has been configured in the controller of the heat exchanger loop and the second one is from 6pm to 8am when heating is not required by administrative and public school buildings.

Figure 2: Heat flow rates in Hotels

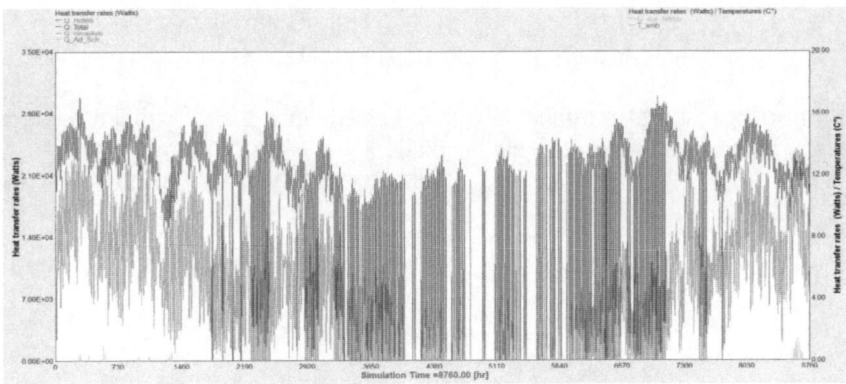

Figure 3: heat flow rates in hospitals

Figure 4: heat flow rates in administrative and public schools

4.2. Heating temperature

The air outside temperature (Tair_in), the air outlet temperature from the heat exchanger (Tair_out) and the final air outlet temperature provided by the whole system (Tair_Final), are plotted in figures 5, 6 and 6, respectively, for hotels, hospitals and administrative and public school spaces. The solar heating system increases the air from the ambient temperature (Tair_in) to the comfort temperature based on the hourly irradiation during the year. It can be seen that the final air outlet temperature is equal to 20°C when the ambient temperature is above the comfort temperature, however the solar heating system doesn't work when the ambient temperature is more than 20°C which means that the air temperature is equal to the ambient temperature or less if the place is equipped by a cooling system.

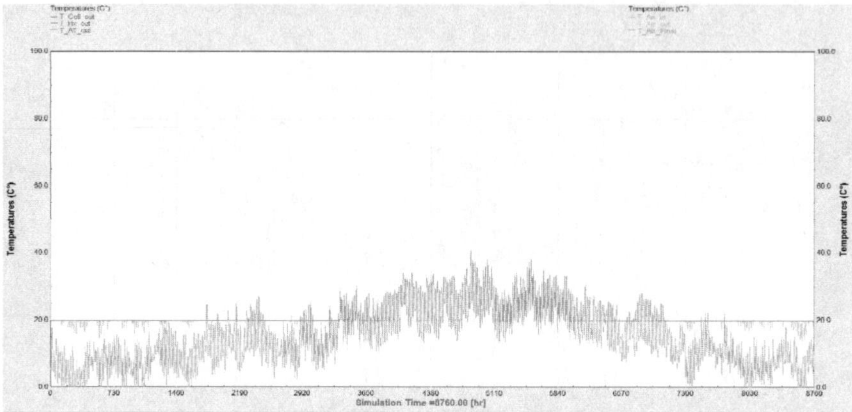

Figure 5: Air temperatures in hotels

Figures 5 and 6 illustrate that the additional heater is needed when the air ambient temperature is less than 1°C and -1°C, respectively, for hotels and hospitals.

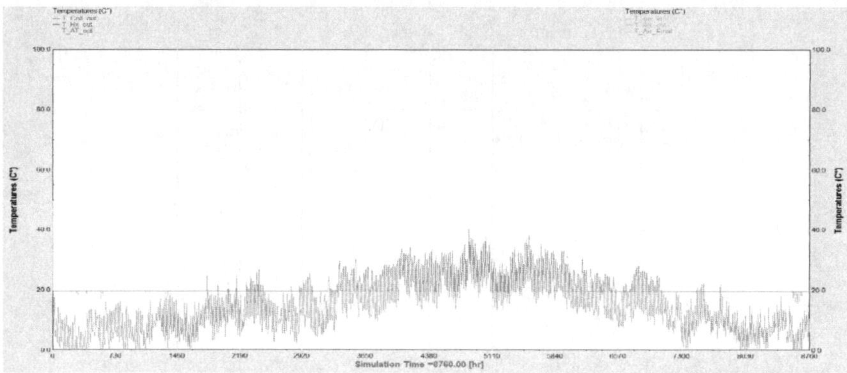

Figure 6: Air temperatures in hospitals

As it can be seen in figure 7, the additional heater doesn't participate in heat generation during the year in administrative and public school buildings thanks to the enough heat produced by the solar loop. Outside the required heating period determined by the control function, the air temperature is equal to the air ambient temperature.

IOP Conf. Series: Materials Science and Engineering **353** (2018) 012004 doi:10.1088/1757-899X/353/1/012004

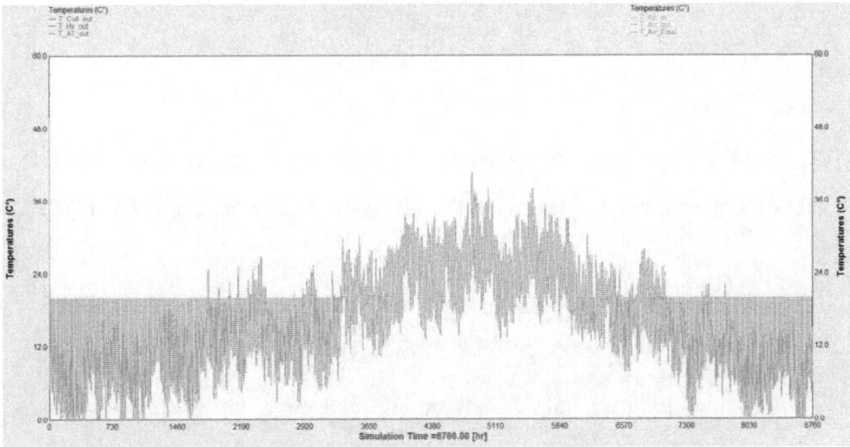

Figure 7: Air temperatures in administrative and public schools

5. Conclusion

Solar air heating is a solar thermal technology where radiations are captured and then transformed by a conversion system to heat the air. It is a renewable energy technology used for heat and cooling generation whether in residential or tertiary sectors. In the present study, the heating system has been designed to ensure a comfort temperature of 20°C and to compensate the thermal heat losses inside a volume of 750 m^3 in tertiary sector. Simulation results have shown that the system can compensate the thermal losses and thus ensure the comfort temperature for the different building categories of the tertiary sector. The sizing of the system differs from one case to another, it depends on the required thermal power and the considered heating period. In this case, the maximal estimated thermal power is around 21 kW, 29 kW and 32 kW based on a solar collector areas of 5 m^2, 7 m^2 and 10 m^2, respectively, for hotels, hospitals, administrative and public-school spaces. This calculation is based on the minimal thermal transmittance recommended by the Moroccan regulation using the TRNSYS environment.

References

[1] Agence Nationale pour le développement des Energies Renouvelables et de l'Efficacité Energétique, *Règlement thermique de construction au Maroc*, ADEREE, 2014.
[2] Daniel Quénard, Buildings: *The new energy nexus*, In Comptes Rendus Physique, Volume 18, Issues 7–8, 2017, Pages 415-427.
[3] Centre Scientifique et technique de la construction, Performance énergétique des bâtiments, CSTC-Contact – N° 9 – Mars 2006.
[4] Bernier Jacques, *La pompe à chaleur*, Editions Parisiennes, 2004.
[5] Gina Penu, *La thermique du bâtiment en 36 fiches-outils*, Dunod, Paris, 2013.
[6] Bououd Mahmoud et Abdellah Mechaqrane. *Concentration Solar Dryer Water-to-Air Heat Exchanger: Modeling and Parametric Studies*. International Journal of Hydrogen Energy 42, n° 13 (mars 2017): 8631-43.

IOP Conf. Series: Materials Science and Engineering **353** (2018) 012005 doi:10.1088/1757-899X/353/1/012005

Simulation of floor heating in a combined solar-biomass system integrated in a public bathhouse located in Marrakech

M KRAROUCH [1*], H HAMDI [1], S LAMGHARI [2], A OUTZOURHIT [2]

[1] Energy Process Research Team, LMFE (URAC27-CNRST)
[2] LN2E, Department of Physics
Faculty of Science Semlalia Cadi Ayyad University, Marrakech, Morocco.

E-mail*: mohamed.krarouch@edu.uca.ma

Abstract. This study was conducted in the framework of the HYBRID-BATH project aiming at improving the energy efficiency of traditional Hammams (Turkish baths) and the reduction of the use of wood energy and therefore of greenhouse gases emissions. The present work focuses on the energetic performance of a two-room Hammam located in Marrakech. The rooms were heated by the ground using a hybrid system Micro-CSP/ biomass boiler. The dynamic simulation of the system (Hammam coupled with the hybrid system Micro-CSP/ biomass boiler) was conducted using TRNSYS18 software. The parametric study was performed on a Typical Meteorological Year data (TMY). This study is devoted to presenting the results of the dynamic simulation of a part of the Hammam investigated, in order to optimize the underfloor heating system. The models and the results of the simulations will be validated by comparisons with experimental results. The main objective is to optimize the operation of such system and to improve its performance.

1. Introduction

The number of traditional Hammams in Morocco is on the order of 10,000 units. A traditional Hammam consumes on average 1.5 ton of wood and between 60 and 120 m^3 of water per day. This significant consumption is mainly due to the low performance of the systems used for heating that generate considerable losses and contribute to over-exploitation of forests. Any approach to reducing this consumption will contribute to the preservation of these two resources that are more and more stressed in Morocco. In this context, it becomes more and more urgent to put in place innovative solutions for the Hammams to limit their impact on the environment.

In this work, we focus on developing other techniques to produce heat and hot fluids based on solar and biomass energy technologies, as Morocco has significant solar resources of about 5 $kWh.m^{-2}.day^{-1}$ and over 3,000 hours of the sunshine per year [1] and undeniable biomass from various agricultural residues. In particular, the Parabolic Trough collector (PTC) will be hybridized with the biomass boilers in order to provide a reliable heat supply to the various components of the Hammam. The biomass boiler will use local resources (olives pomace, walnuts, almonds and argan shells...).

Indeed, the objectives of the HYBRIDBATH project (Hybrid solar / biomass systems for the production of hot water and heating in traditional Hammams) are two-fold. 1) to improve and standardize the combined solar-biomass heating systems ; 2) to advance the research in the technical interaction between the biomass boiler, the water storage system, the solar thermal collector, and floor heating. The overall aim of the project is to improve the energy efficiency of traditional Hammam (Turkish baths) and to reduce the consumption of wood energy and therefore of emissions of greenhouse gases.

Many studies available in the literature evaluated the contribution of some hybrid solar-biomass heating system through simulations and experiments. In this regard, D. Chasapiset al. [2] presented a complete cycle of the operational results of a thermal solar-biomass hybrid system for space heating in Greece. They found that the results are promising with respect to the technology and energy requirements: the daily coverage of the heating from the solar collectors often exceeded 40%. The CO2 emissions avoided by the use of this system from November 2005 until the end of April 2006 have been calculated as 1.016 tons. A similar study based on a dynamic simulation by TRNSYS was conducted by Florian Stift et al. [3] who studied the combined biomass-solar thermal system for heating and domestic hot water supply, designed for small residential applications. The system consists of a thermal storage with an integrated pellet burner and a heat exchanger to feed solar energy from a small collector field into the storage. They have reduced the use of biomass by 17% and an increase of 11% of the solar energy supply could be achieved.

A detailed study of combined solar and pellet heating systems have been carried out by Fiedler [4]. In the study 4 solar and pellet heating systems were studied with the help of annual dynamic simulation. Two of the systems comprised a pellet stove and two systems were solar combisystems; one with a store integrated pellet burner, the other with a separate pellet boiler. The biomass stove and boiler model developed by Nordlander were used in these studies [5].

Other studies focusing on the technique of solar heated floor were conducted in the literature. Martinopoulos and Tsalikis [6], studied the active solar heating systems for energy efficient buildings. They performed a technical and economic evaluation of a solar space heating system, installed in a house of 88 m² surface. The analysis was performed for each of the four climate zones designated in the thermal regulation in Greece. They found that the solar system covers at least 45% of the total heating load whereas the recovery period was less than 4.5 years with an annual reduction of more than 50 tons of CO2 in the worst case. Sobhy et al [7] worked on improving the energy performance and thermal behavior of a solar system for heating a single Family Hammam through a heated floor in Marrakech. Two configurations with or without thermal storage, were compared. The dynamic simulation with the TRNSYS software was used to optimize the solar floor heating system.

Furthermore, the experimental study on floor heating was made by Kharchi et al. [8] who studied the thermal behavior of a system with direct solar floor in Bouzaréah (Algeria). Twenty thermocouples were installed at various points of the slab as well as in the inlet and in the outlet of the slab. The authors found DSF (direct solar floor) lowered the temperature of the zone by 1 to 2 °C. In addition, the DSF avoided the phenomenon of cold zone, and the heat was evenly distributed. The dynamic simulation of solar heating floor system by TRNSYS was done by Sobhan-sarbandi [9]. He carried out a comparative study of the effectiveness of two type of solar collectors to heat the floor of a house in Gazimagusa in northern Cyprus under the same conditions during the five coldest days of winter. The first is composed of solar flats collectors in series with a total surface area of 8 m². The second was a compound parabolic collector (CPC) of an area of 2 m². The results showed that the CPC collector provided an equivalent performance to all of the flat collectors in series, and it was able to maintain the desired comfort floor temperature of 25 °C.

The aim of this work is to analyze the energetic performance of Hammam (Turkish bath) situated in Marrakech heated by underfloor heating system. The Dynamic simulation of the Hammam coupled with the hybrid Micro-CSP/ biomass boiler system was conducted using TRNSYS software.

2. Description of the hybrid heating system installed

2.1. Parabolic trough collectors

The increase in the use of solar collectors in recent years for domestic hot water preparation has shown that solar heating systems are a mature and reliable technology. Motivated by the confirmed success of these systems for hot water production, an increasing number of home builders are considering solar energy for space heating as well.

In this work we will use the Parabolic Though Collectors (PTC) (TREVELLI, Italy) to pre-heat the water for the Hammam. The PTC collects beam radiation, within a critical angle called the half-acceptance angle and focusses it on its focal line where receiver is placed. Technical characteristics of Parabolic Trough collectors used in this study are presented in Table 1.

Table 1. Technical characteristics of parabolic troughs.

IOP Conf. Series: Materials Science and Engineering **353** (2018) 012005 doi:10.1088/1757-899X/353/1/012005

Number in series (PTC)	*6*
Area	*8,25 m² * 6*
Fluid used	*Eau + glycol*
Maximum Temperature	*250°C (programmable)*
Tracking	*Mon-axial*
Peak Power (at DNI 1000 W/ m²)	*6 Kw*
Life expectancy	*20 years*

2.2. Biomass boilers

The Biomass boilers (Klorina 22, TATANO, Italy) used in this work run at an efficiency of 89 – 91% compared to 28 to 32 % for traditional ones used in public Hammams, with a thermal power output of 2*115 kW. The Biomass feed and air flow can be readily controlled. We plan to use a renewable biomass as feed that have a minimal impact on the environment, such as:

- Pellets,
- Olive pommace,
- Almond shells,
- Argane shells
- Nut shells, etc.

2.3. Floor heating

Under-floor heating is a method heat is provided through the floor. The floor heating system is installed in the three rooms (hot, warm1 and warm2) of each compartment. It covers a surface of 29.6 m² in the hot and warm rooms, and 33 m² in the second warm room, as shown by the Figure 1.

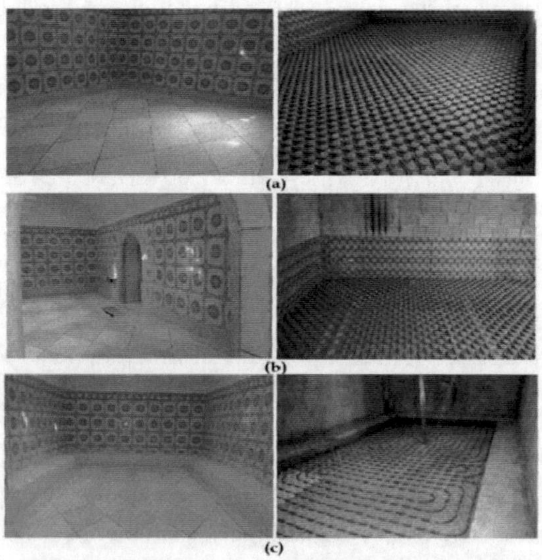

Figure 1.Three different underfloor heating systems, (a) hot room, (b) the first warm room, (c) the 2nd warm room.

3. Description of the hybrid system studied with TRNSYS

The essential idea is to couple the heat emission device used for space heating to both the boiler biomass and the solar collectors in such a way that the users of Hammam can benefit from the heating floor comfort throughout all the day. The system is modelled in TRNSYS, and the model includes collectors, collector loop, storage, boiler biomass, building (floor heating), pumps, and control system. A schematic diagram of the system is shown in Figure 2. As shown in this Figure, the buffer tank for

IOP Conf. Series: Materials Science and Engineering **353** (2018) 012005 doi:10.1088/1757-899X/353/1/012005

space heating is heated by the heat exchanger of the parabolic trough collector. When solar energy is not adequate, the temperature in the top of the buffer tank drops below the set point and the system will be backed up by the biomass boiler. Furthermore, the boiler will be operated at the maximum set temperature, which is 80 °C.

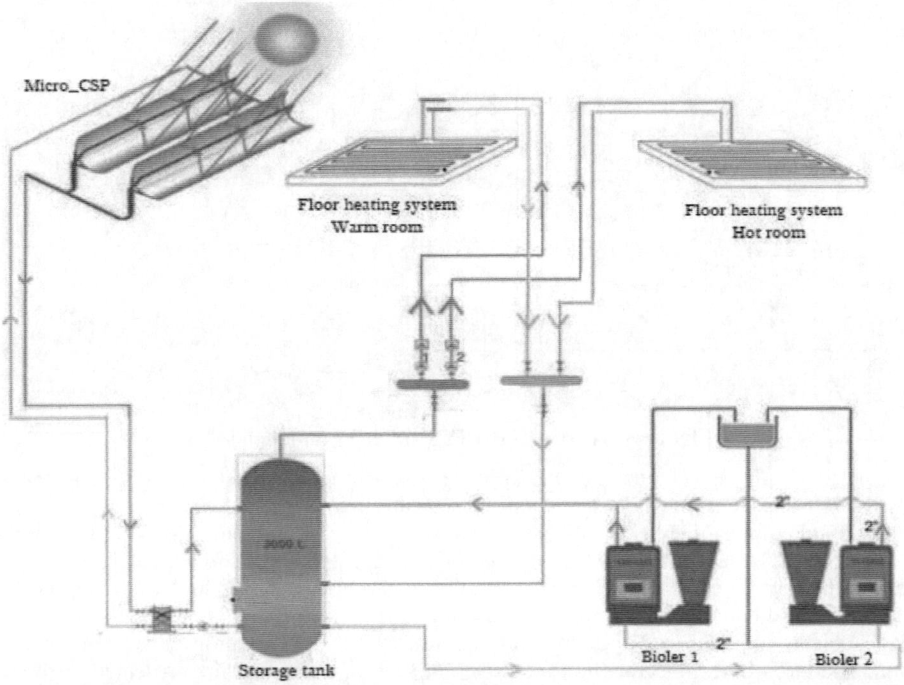

Figure 2. Hydraulic schematic for the simulation of the solar combisystem

4. Dynamic simulation

The thermal behavior of the part of Hammam studied with heated floor coupled to the hybrid solar/biomass system is modeled through the dynamic simulation software TRNSYS 18 using the model of building (Types 56) [10] as shown in Figure 3.

Transient Systems Simulations (TRNSYS) software is a complete and extensible simulation environment for the transient simulation of thermal systems including multi-zone buildings. It is widely used around the world by engineers and researchers to investigate alternative energy applications, from some simple systems such as domestic hot water system to the design and simulation of buildings and their equipment, including residents behavior, control strategies, alternative energy systems (wind, solar, etc.) [11].

The simulation of the building was carried out using standardized meteorological data (TMY) from Meteonorm [12]. TMY reads the weather information regularly, and checks the solar radiation data to find tilted surface radiation and angle of incidence for an arbitrary number of surfaces. Marrakech climate conditions are used for simulation. The weather data are given on hourly basis.

5. Results and discussion

The studied system is located in Marrakech, Morocco (31°38'N latitude, longitude 8°03'W, altitude 426 m), 200 km from the Atlantic coast. The climate of Marrakech is characterized by approximately cold winters and very hot summers, as well significant differences between day and night temperatures. The climate is classified as semi-arid with 3000 h/year sunshine [1]. Figure 4. shows the average hourly temperature of the ambient air for a typical meteorological year in Marrakech.

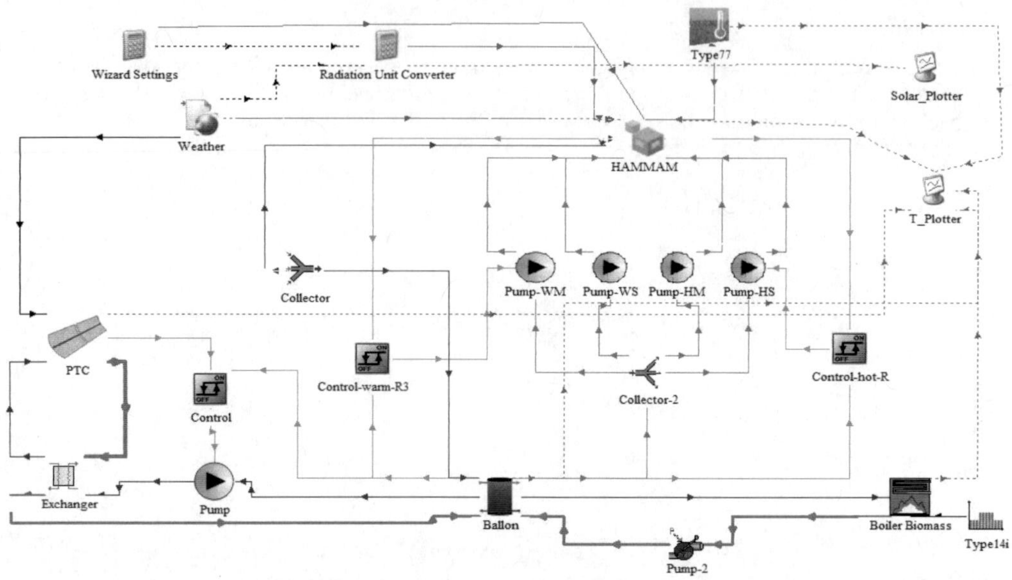

Figure 3. Thermal TRNSYS simulation model

The main objective of this work is to study the influence of solar combisystem on the thermal behavior of the part of Hammam during the winter period when heating is needed, in order to achieve the comfort temperature for a Public bathhouses. The simulations were done during the month of December because it is one of the coldest months of the year. As shown in Figure 4. January and December are the coldest months with an average air temperature of 12 °C and 13 °C respectively. The results are obtained during the last week of December.
The hourly ambient temperature (Ta) of Marrakesh and also the total radiation, direct horizontal and direct on the aperture of the collector from the 24th to the 31st of December are presented in figures 5 and 6. It can be observed that, during the night, Ta can be as low as 16-18 °C, while during the day it is between 28 and 33 ° C. The total solar irradiation in Marrakech on a horizontal plane ranging from 3 to 3.6 kWh m^{-2} day^{-1}.

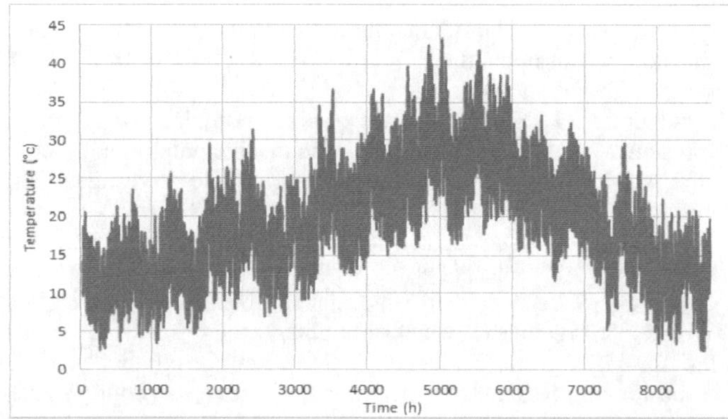

Figure 4. Outside temperature variation during year in Marrakech (TMY data)

A reasonably accurate knowledge of the solar radiation is necessary for solar energy systems such as solar thermal. In this regard, it is important to be able to predict the angle between the sun radiation and a vector normal to the aperture or surface of the collector. In this study we used the two axis-tracking mode for the PTC, because the angle of incidence is equal to zero in this case. Therefore the maximum amount of the direct solar radiation is collected as the collector aperture points directly towards the sun.

IOP Conf. Series: Materials Science and Engineering **353** (2018) 012005 doi:10.1088/1757-899X/353/1/012005

The results obtained in Figure 6. show that during the day, the temperature of the water flow increases as solar energy is absorbed by the solar collectors. The outlet fluid temperature for the PTC is between 52-190 °C. During night time water will be stored in storage tank.

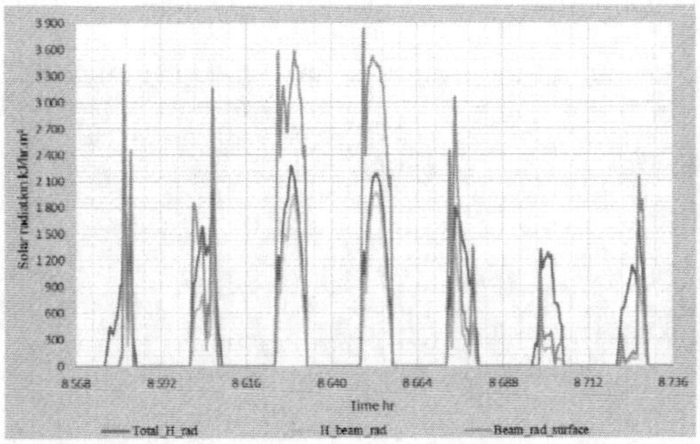

Figure 5. solar radiation -24th until 31st of December

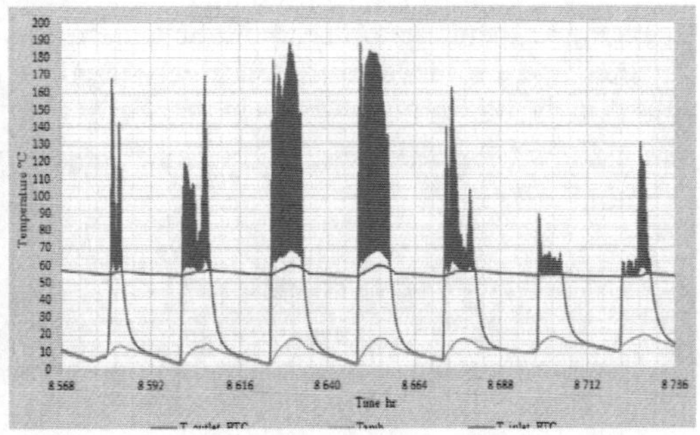

Figure 6. the hourly variation of inlet and outlet water flow temperature - PTC collector -24th until 31st of December

6. Heating the Hammam by the combined system in cold winter period.

6.1. Building temperature evolution

The studied Hammam composed of two Rooms (Hot and Warm), each one is a monozone building (6×3×3 m^3). The maximum performances from the combined system is achieved by use of 8,25 m^2 of parabolic trough collector, a 100 kg/hr of collector mass flow rate and a 3000 l of hot water storage tank. The heating system is embedded in the floor and some parts of the walls at 0.8 m height, the mass flow rate of water circulated through the radiant floor $\dot{m}$=300 kg/hr, and the simulation time step Δt=0.125 hr.

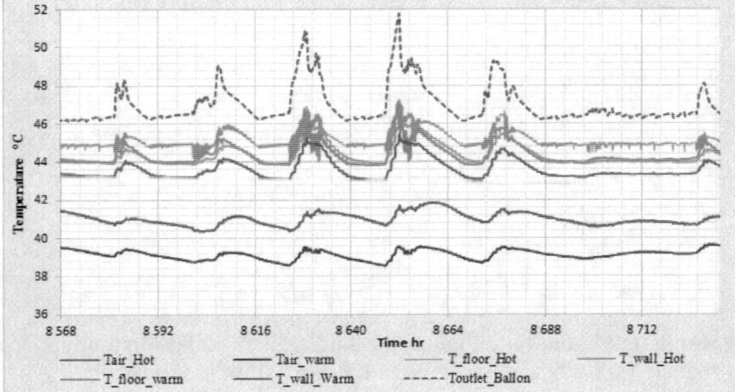

Figure 7. Evolution of different temperatures inside the hot and warm rooms - 24 th until 31st December.

The results show that during the whole day the boiler is turned off and only heat from the solar collectors is used. At night or during a cloudy day; if the temperature in the hot water tank gets below the set temperature (60 °C), the controller will turn on the biomass boiler to keep the top of the tank at the needed temperature.

The analysis of the Figure 7. shows that the operating temperature does not exceed 43 °C in the hot Room and 40 °C in the warm room. In addition, the temperatures of the floor and wall surfaces in the warm and hot rooms are between 43 and 46 ° C. Further, this range represents the temperature of comfort inside the traditional Hammams in Marrakech.

6.2. The effect of tubing materials on the floor temperature

The thermal properties of under floor heating tubing can play an important role in the efficiency of the heating system. In this study, a comparison between the two tubing materials commonly used as floor heating coil, namely copper and crosslinked polyethylene (PEX) was carried out. The thermal conductivity of copper tubes and PEX are 328 $W.m^{-1}.K^{-1}$ and 0.45 $W.m^{-1}.K^{-1}$ respectively. The figures below present the temperature of the heated floor surface with copper and PEX pipes and the operative temperatures of the two rooms in the presence of water flow regulation.
Figure 8 and 9 shows the difference between the temperature of the heated floor with copper and PEX tubing in the hot room as well as in the Warm room.

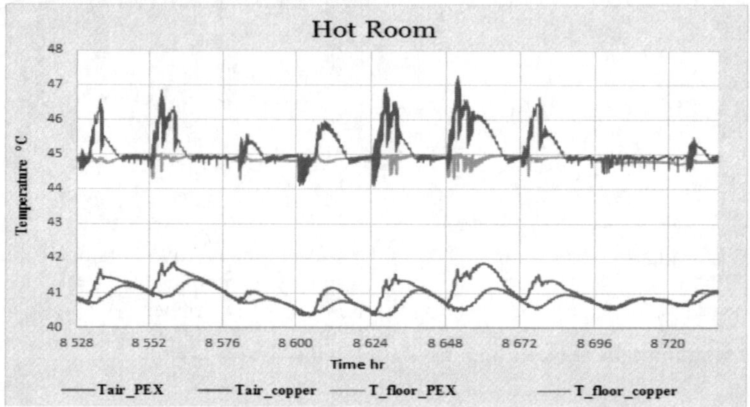

Figure 8. The operative room temperature and floor heating surface temperature for copper and PEX tubes in warm room-22th until 31st of December

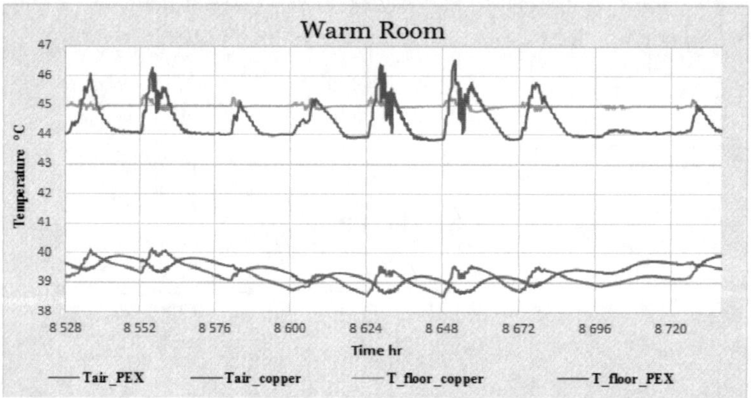

Figure 9. The operative room temperature and floor heating surface temperature for copper and PEX tubes in warm room-22th until 31st of December

As shows in Figure 8, the temperature of the heated floor surface has the same behavior, the maximum surface temperature difference between the two serpentine types is 2 °C. For the warm room, as shows in Figure 9, the maximum surface temperature difference between the two serpentine types is on the order of 1 °C.
From these results, we note that, although copper tubing has a higher thermal conductivity compared to the PEX one, it has not a significant effect on the heating floor surface temperature. However the cost of Copper tubing is higher than the cost of PEX tubing. The economic solution is to use PEX tubing as a tubing material.

6.3. The effect of the distance between tubes on floor surface temperature

Figure.10 shows the effect of changing the distance (dx) between the tubes on the floor surface temperatures of the hot and warm rooms in the last week of December. From the figures below we note that the surface temperature of the floors increases with decreasing dx. We also note that for the hot room, the floor surface temperature for dx = 7 cm is between 44 to 47,5 °C and for dx= 16 cm it is between 44,5 to 45,5°C, furthermore, the floor surface temperature of warm room for dx = 7 cm is between 44 to 46,5°C and for dx= 16 cm is between 43,5 to 45°C. We found that with the controllers that are used to regulate mass flow rates, with an even smaller dx, it is expected that the maximum floor surface temperature would exceed the thermal comfort criteria of the Hammam (46 °C) for both hot and warm room.

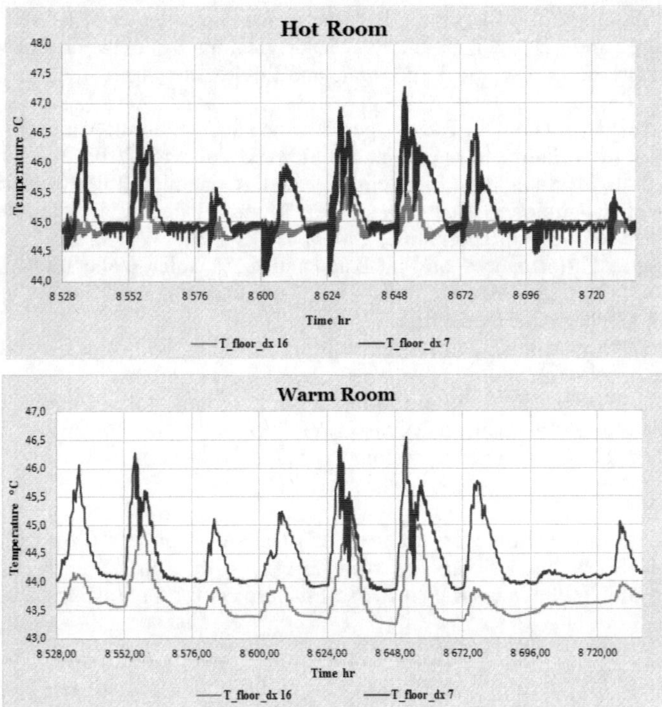

Figure 10. Floor surface temperature in the hot and warm room with different distance between tubes-22th until 31st of December.

7. Conclusion

In this work, we modelled a combined solar/biomass system for heating two rooms of the Hammam located in the city of Marrakech during the last week of December. Dynamic simulation via TRNSYS software was used in order to optimize the floor heating system. The biomass boiler is required to overcome the shortcomings of solar energy (intermittency), in order to keep the operative temperature at the zone of comfort of a Hammam throughout the day.

The influence of pipe material on floor surface temperature was investigated. It is shown that there is no significant difference of floor heating either temperature when copper or PEX tubing are used. The optimum solution is therefore to use PEX tubing as pipe material. In addition, we studied the effect changing the distance between the tubes on the floor surface temperatures for the hot and warm rooms. The next steps concern to study a combined biomass-solar system for domestic hot water production and space heating, of a full global model of Hammam coupling of these different models under the environment of TRNSYS 18. The validation of the model will be realized through the comparison with the experimental measurements, which will be carried out. The main objective is to optimize the operation of such system and to improve its performance.

8. References

IOP Conf. Series: Materials Science and Engineering **353** (2018) 012005 doi:10.1088/1757-899X/353/1/012005

[1] National Agency for the Development of Renewable Energies and Energy Efficiency, ADEREE, Morocco, 2015, www.aderee.ma.
[2] D. Chasapis, V. Drosou, I. Papamechael, A. Aidonis , R. Blanchard, Monitoring and operational results of a hybrid solar-biomass heating system, Renewable Energy 33 (2008) 1759–1767.
[3] F Stift, M Hartl, T Ferhatbegović , S Aigenbauer, A Simetzberger, Model based optimization of a combined biomass-solar thermal system, Energy Procedia 48 (2014) 681 – 688.
[4] Fiedler, F., 2006. Combined Solar and Pellet Heating Systems: Study of Energy Use and CO-Emissions. PhD Thesis, Mälardalen University, Department of Public Technology.
[5] Nordlander, S., 2003. TRNSYS Model for Type 210 - Pellet Stove with Liquid Heat Exchanger - Documentation of Model and Parameter Identification. Report ISRN DU-SERC--79--SE, Högskolan Dalarna, Solar Energy Research Center SERC, Borlänge.
[6] G. Martinopoulos, and G. Tsalikis, "Active Solar Heating Systems for Energy Efficient Buildings in Greece: A Technical Economic and Environmental Evaluation", Energy and Building,Vol. 68, 2014, pp.130-137.
[7] I Sobhy, A Brakez, B Benhamou, Energy performance and economic study of a solar floor heating system for a Hammam Energy and Buildings, Vol. 141 (2017) 247–261.
[8] R. Kharchi, N. Aịt Messaoudčne, M. Belhamel and A. Hamid,"Etude Expérimentale du Comportement Thermique d" un Plancher Solaire Direct", Revue des Energies Renouvelables: Journées de Thermique, 2001, pp. 115-119.
[9] S. Sobhansarbandi, Performances of Flat-Plate and CPC Solar Collectors in Underfloor Heating Systems, Eastern Mediterranean University Gazimağusa, North Cyprus, Master of Science in Mechanical Engineering, February, 2013.
[10] TRNSYS 18, reference manual, Multizone Building modeling with type 56 and TRNBuild, 2017-02, Madison,WI, USA.
[11] S. Klein, et al, "A Transient System Simulation Program, Solar Energy Laboratory", University of Wisconsin-Madison, Madison,WI, USA.
[12] Meteonorm V7.0.22.8. www.meteonorm.com.

Acknowledgments
This study is carried out in the framework of the HYBRIDBATH project financially supported by IRESEN (Institute of Research on Solar and New Energy -Morocco). We are also grateful for Watersolar Maroc and Buildit, the industrial partners of the project for their technical support.

IOP Conf. Series: Materials Science and Engineering **353** (2018) 012006 doi:10.1088/1757-899X/353/1/012006

Statistical analysis of global horizontal solar irradiation GHI in Fez city, Morocco

Z BOUNOUA[1] and A MECHAQRANE[1]

[1] Laboratory of Renewable Energies and Intelligent System Electrical Engineering department, Faculty of Sciences and Technologies, Sidi Mohamed Ben Abdellah University of Fez PO. Box 2202 Fez Morocco

zineb.bounoua@usmba.ac.ma

Abstract. An accurate knowledge of the solar energy reaching the ground is necessary for sizing and optimizing the performances of solar installations. This paper describes a statistical analysis of the global horizontal solar irradiation (GHI) at Fez city, Morocco. For better reliability, we have first applied a set of check procedures to test the quality of hourly GHI measurements. We then eliminate the erroneous values which are generally due to measurement or the cosine effect errors. Statistical analysis show that the annual mean daily values of GHI is of approximately 5 kWh/m²/day. Daily monthly mean values and other parameter are also calculated.

1. Introduction

In the field of solar energy, it is important to know the amount of solar radiation incident on the ground at every moment, in order to properly manage solar power plants. It is therefore necessary to take measurements of all solar components at different time scales and to create databases of meteorological information. The reliability of these solar data is not always guaranteed, because the measurements of the pyranometer can be affected by several errors due to technical problems encountered during the measurement [1, 2]. A quality control of the available data has been achieved in a previous work [3].

In this paper, we will statistically analyse temporal variations of global horizontal irradiation (GHI) measured for six years. This study will subsequently help researchers to evaluate the amount of energy available in Fez city to use it in other research projects.

This paper is organized as follows: the second part is dedicated to the description of the database used in this study. The third part represents the necessary processing performed on the data before exploitation. The results of the statistical analysis are shown in the fourth part.

2. Database

In this study, hourly measurements of GHI were performed using a Kip & Zonen pyranometer (model CM11). This instrument is part of a meteorological station placed on the roof of the Faculty of Sciences and Technologies of Fez (Latitude: 33° 59' 58" N, longitude: 4° 59' 22" W and altitude: 450 m). The measurement period of the database used is defined between 01/01/2009 01:00:00 GMT to 10/06/2015 at 24:00:00 GMT. Figure 1 shows the hourly variations of GHI for the whole considered period.

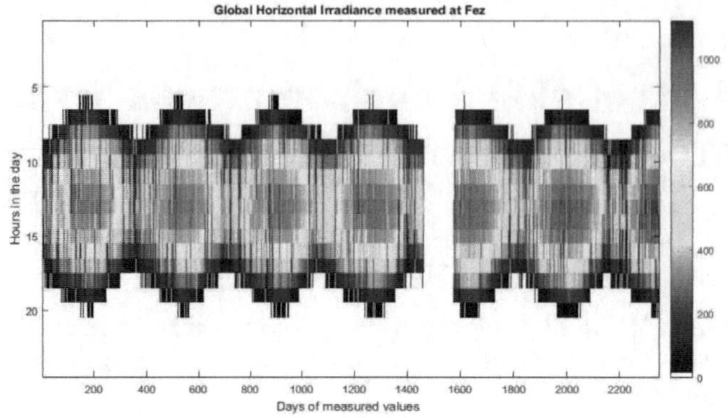

Figure 1. Hourly variations of the measured GHI for the whole considered period.

According to this figure, we notice the existence of a data hole in GHI measurements in the beginning of the year 2013. This is due to a problem in the connection cable of the pyranometer to the data logger.

3. Data processing

Before beginning the statistical analysis, we performed several calculation and correction techniques represented below:

- The Solar Position Algorithm (SPA) [4] was chosen to calculate the solar position at each time. SPA was developed at the National Renewable Energy Laboratory (NREL) of the U.S. Department of Energy's to calculate the sun's zenith and azimuth angles and other related parameters with an unmatched low uncertainty of +/- 0.0003 degrees on a valid period of years from -2000 to 6000 [5].
- A set of quality control tests was applied to achieve an acceptable quality of GHI measurements [3]. For this purpose we relied on several previous works [2, 6, 7], and adapted modifications to our data. After applying this quality control procedure, we have 19017 hourly of GHI measures that passed all tests successfully and 6896 that have been eliminated after those tests.

4. Statistical analysis and results

In this part, we will analyse our data and present the distribution of GHI between the year 2009 and 2014. To carry out this analysis, we calculated the hourly mean, the monthly daily mean and the annual daily mean of GHI.

4.1. The mean hourly values of GHI:

Figure 2 shows the GHI mean hourly values calculated from the whole hourly values measured between 2009 and 2014. We can see a normal shape for hourly values and to be more precise we have added the error bar, which can come from the pyranometer precision (2%).

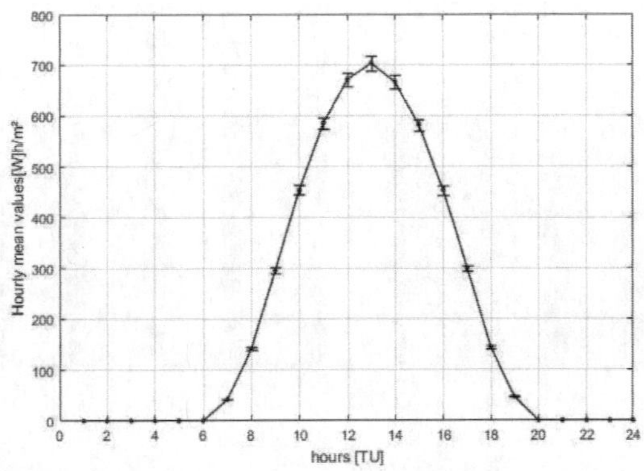

Figure 2. Mean hourly values of GHI in the city of Fez.

4.2. The daily values of GHI:

Figure 3 shows the variations of daily values. We note the presence of several missing data because there are few days that contain one or more missing hours, so to do this analyse we first eliminate those incomplete days, we consider only the day without any missed hourly value.

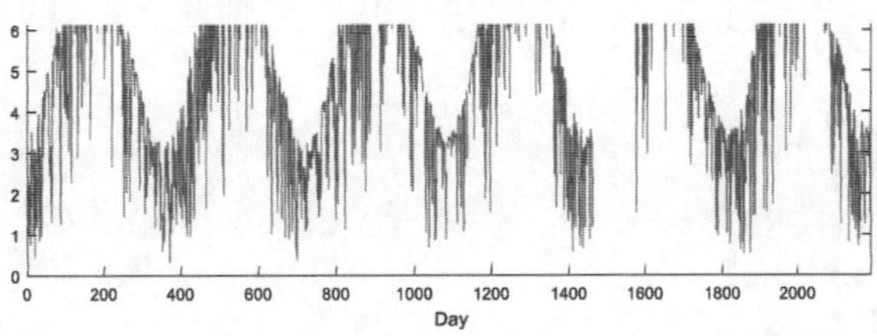

Figure 3. The daily values of GHI in the city of Fez.

4.3. The monthly daily mean values of GHI:

Figure 4 shows the monthly daily mean values (kWh/m²/day). We note that the minimal value is of 2.39 kWh/m²/day and it observed in December month. We can also see that the period from May to August is well sunned: GHI exceeds 6 kWh/m²/day.

IOP Conf. Series: Materials Science and Engineering **353** (2018) 012006 doi:10.1088/1757-899X/353/1/012006

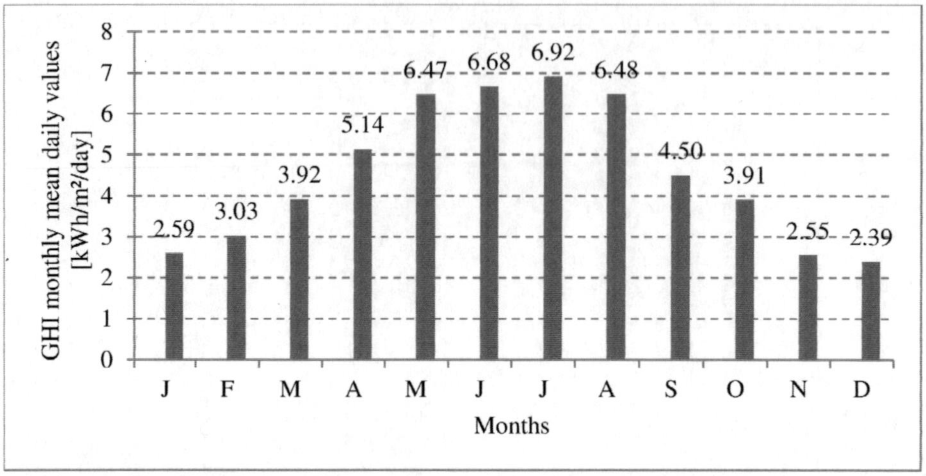

Figure 4. The monthly daily mean values of GHI in the city of Fez.

4.4. The annual daily mean values of GHI:
Table 1 shows the annual daily of GHI calculated according to the number of full days corresponding to each year. We note that the annual daily mean of GHI varies slightly. The minimal observed value is 4.64 kWh/m²/day.

Table 1. The values of the annual daily mean of GHI between 2009 and 2014

Year	GHI(kWh/m²/day)	Number of full days
2009	4.80	362
2010	4.64	357
2011	4.96	357
2012	5.17	358
2013	5.72	252
2014	5.37	364

Only the year 2013 is not so longer because of the problem in the cable of the pyranometer, its annual value is calculated only for the 252 full days. By omitting this year, the mean annual daily value is of 4.99 kWh/m²/day.

5. Conclusion

The objective of this work is to analyse and evaluate the global horizontal solar irradiation measurements taken in Fez city between the year 2009 and 2014. We statistically analyse these data by calculating and representing the monthly averages, daily monthly averages and daily annual average values of GHI. We found that the minimal monthly mean daily value is of 2.39 kWh/m²/day and it observed in December month. The yearly mean GHI daily value is of approximately 5 kWh/m²/day. This shows that the region of Fez city is a good-sunned site.

6. References

[1] Moradi I 2009 Quality control of global solar radiation using sunshine duration hours *Energy* 34.1: 1-6.

[2] Younes S, Claywell R and Muneer T 2005 Quality control of solar radiation data: present status and proposed new approaches Energy 30:1533–49.

[3] Z.Bounoua, A.Mechaqrane, Assessment of a quality control procedure of hourly solar irradiations at Fez city, Morocco, The International Conference on Renewable Energies and Energy Efficiency FST Fez, November 8 & 9, 2017.

IOP Conf. Series: Materials Science and Engineering **353** (2018) 012006 doi:10.1088/1757-899X/353/1/012006

[4] Reda I. and Andreas A. 2003 Solar position algorithm for solar radiation applications: Technical Report, National Renewable Energy Laboratory, Golden, Co, USA. Revised version: January 2008.

[5] Blanc P and Wald L 2012 The SG2 algorithm for a fast and accurate computation of the position of the Sun for multi-decadal time period. Solar Energy, 86(10) 3072-3083.

[6] Ruiz-Arias J.A., Alsamamra H., Tovar-Pescador J and Pozo-Vazquez D 2010 Proposal of a regressive model for the hourly diffuse solar radiation under all sky conditions Energy Convers Manag, 51:881–893.

[7] Ihya B , Mechaqrane A , Tadili R and Bargach M N 2014 Prediction of hourly and daily diffuse solar fraction in the city of Fez (Morocco), Theoretical and Applied Climatology, 1-13.

IOP Conf. Series: Materials Science and Engineering **353** (2018) 012007 doi:10.1088/1757-899X/353/1/012007

Cooling of a dwelling by nocturnal radiation

Othmane Fahim[1], Naoual Belouaggadia[1], Mohamed Taqi[1] and Chérifa Abid[2]

[1]: Laboratory of Engineering and Materials (LIMAT), Ben M'Sik Faculty of Sciences, Hassan II University of Casablanca, B.P. 7955 Sidi Othmane, Casablanca, Morocco.

[2]: University Polytechnic School of Marseille, Department of Energy Mechanics, University of Provence.

E-mail: othmanefahim@gmail.com

Abstract - Atmospheric transparency in the infrared, responsible for night cooling, is exploited to obtain a cooling effect. Radiative cooling to the night sky is based on the principle of infrared radiation heat loss from a surface to a body at a lower temperature. The use of the emissivity equation allowed us to evaluate its variation as a function of wavelength and temperature. A comparison of the temperature variation was made between granite and the materials most often used in the manufacture of radiant panels of hybrid systems. The results show that the temperature of Tedlar-based plates or plastics considerably decreases, and, therefore are rather promising.

1. Introduction

Today, because of the expected shortage of energy supply and air pollution problems, energy conservation is really taken into account not only in the industrial sector, but also at the residential level. The energy required for heating and cooling buildings accounts for about 6.7% of total global energy and most of the cooling systems used are huge energy consumers. The use of passive cooling systems can contribute to a saving of 2.35% of global energy needs [1]. Night-sky cooling to the sky is based on the principle of radiation heat loss from one surface to another at a lower temperature. The roofs of buildings radiate heat, day and night, at an average rate of 75 W / m² during the day [2]. The use of the window of transparency of the atmosphere, in the wavelength range of the average infrared (8-14 µm), allows on a clear day, to selective materials whose surface is strongly emitting in the infrared to strongly cool. In this work, we propose to show that we can intensify this effect by a good choice of materials. We perform a comparative study between the performance of the proposed material and the materials most often used in the manufacture of this type of surfaces.

2. Determination of the cooling power

To determine the power of radiative cooling, we must establish the energy balance on a surface. From Fig. 1, we note that if a surface is in equilibrium with its environment, the energy balance of this surface can be expressed by:

$$\phi_{in} = \phi_{sun}(T_{sun}) + \phi_{atm}(T_{amb}) = G_{sun}(T_{sun})\alpha_e T_e + \varepsilon_{sky}(T_e)\alpha_e T_e^4 \quad (1)$$

$$\phi_{out} = \phi_{emittre}(T_e) + \phi_{cv} = \varepsilon_e(T_e)\sigma T_e^4 + h_c(T_{amb} - T_e) \quad (2)$$

$$\phi_{cooling} = \phi_{out} + \phi_{in} \quad (3)$$

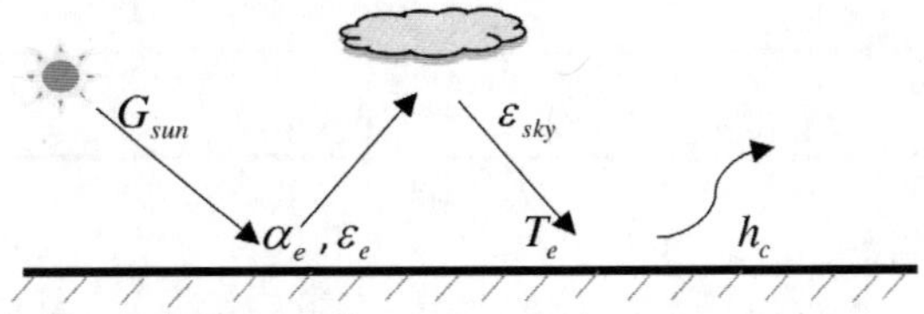

Figure 1. Diagram explaining the energy balance of a surface

Here, φ is the heat flux transferred to or from the surface of the emitter, and G is the irradiation of the sun. The thermal flux transferred by natural or forced convection (wind) is taken into account in the formulation, with h_c the coefficient of heat transfer by convection. The emission and absorption coefficients are respectively defined by ε and α, which are given here as the total and the average for all the surfaces of the transmitter. In order to correctly model the radiative cooling, we have to express the absorption emission coefficients as a function of the wavelength λ.

One of the situations required for night cooling is a clear sky and a dry air. In a clear sky, without dust or humidity, we could have a nocturnal radiative cooling, i.e. φ_{in} π φ_{out}. However, if there is dust and high humidity during the night, we have: $\varphi_{in} \approx \varphi_{out}$ and we cannot have radiative cooling. We must emphasize that the most important consideration here is the spectral radiative transfer to and from the transmitter at different wavelengths. With the net spectral change of the radiative flux, we can have the radiation necessary for cooling. Therefore, the equations governing radiative cooling must be written considering monochromatic quantities. For this purpose, we consider an emitting surface (Fig. 1) at temperature T_e, with emissivity ε_e (λ,θ) and spectral absorptivity α(λ,θ). In this formulation, we assume that a surface is exposed to a clear sky, is the case effectively at 0 K. Of course, the surface may receive the solar radiation and the atmospheric irradiating corresponding to an ambient temperature T_{amb}. The cooling per unit area of a surface is then expressed by the following relation: $P_{cooling} = P_{in} - P_{out}$ and $P_{out} = P_{rad}(T_e) + P_{cv}$ With:

$$P_{cv} = \phi_{cv} = h_c (T_{amb} - T_e) \tag{4}$$

$$P_{rad} = \phi_e(T_e) = 2\pi \int d\Omega \times \cos\theta \int_0^{\infty} d\lambda \times I_{BB}(T,\lambda) \times \varepsilon(\lambda,\Omega) \tag{5}$$

P_{cv} is the resulting power of the exchange by convection with the atmosphere. P_{rad} The radiation from the surface area per unit of surface, and integrated on the solid angle encompassing the Hemisphere

$$d\Omega = \int_0^{\pi/2} d\theta \times \sin\theta \int_0^{2\pi} d\varphi \tag{6}$$

The radiative intensity of the spectral blackbody is expressed by:

$$I(T,\lambda) = (\frac{C_1}{\lambda^5}) / \left[e^{\frac{C_2}{\lambda T}} - 1 \right] \tag{7}$$

The expression of the coefficients C_1 and C_2 is $C_1 = 2hc^2$ with $C_2 = h\frac{c}{k}$

$$P_{in} = P_{atm} + P_{sun} = \int d\Omega \cos\theta \times \int_0^{\infty} \varepsilon\ (\lambda,\theta) \times \varepsilon_{sky}(\lambda,\theta) \times I(\lambda, T_{skyl}) d\lambda + \int_0^{\infty} \varepsilon\ (\lambda,\theta) \times I(\lambda) d\lambda \tag{8}$$

P_{ent} is the incident solar power absorbed by the surface per unit area. To obtain P_{soliel} and P_{atm}, we use Kirchhoff's law $\alpha_\lambda = \varepsilon_\lambda$ to replace the absorptivity of surface by the emissivity. Particularly, there are many correlations to calculate the emissivity of the sky which depend on the state of the sky clear or cloudy, we will remember them in Table.1

Clear sky		Cloudy sky	
Reference	Correlation	Reference	correlation
[2]	$\varepsilon_{cc} = 1 - 0.261\exp[-7.77\times10^{-4} \times (273 - T_{amb})$	[6]	$\varepsilon_{ciel} = (1 + 0.0496c^{2.45})\varepsilon_{cc}$
[3]	$\varepsilon_{cc} = 0.7 + 5.95\times10^{-5} e_{amb} \exp(1500/T_{amb})$	[7]	$\varepsilon_{ciel} = (1.03 + 0.34c\ )\varepsilon_{cc}$
[4]	$\varepsilon_{cc} = 1.24\times(e_{amb}/T_{amb})^{1/7}$	[6]	$\varepsilon_{ciel} = (1 + 0.22c^{2.75})\varepsilon_{cc}$
[5]	$\varepsilon_{cc} = 0.770 + 0.0038T_{dp}$	[9]	$\varepsilon_{cc} = \varepsilon_{cc} + (1 + 0.0024n + 0.0035n^2 + 0.00028n^3$

Table 1. Formulas for calculating emissivity in a clear and cloudy sky

3. Choice of materials

The emissivity of natural substances depends, in a general way, on their physical-chemical nature, their state of geometrical surfaces (flatness defects, roughness) and varies with the wavelength, the direction of emission and the surface temperature. The monochromatic expression is as follows:

$$\varepsilon_{\lambda}(T) = \frac{\int_0^{\infty} \varepsilon(\lambda,T) M^0_{\lambda,T} d\lambda}{\int_0^{\infty} M^0_{\lambda,T} d\lambda} \tag{9}$$

The systematic study of the properties of the few materials used in the manufacture of radiative panels led us to select four materials: Tedlar, already used by Silvestrini [9], white plastic, copper and granite. The study of the thermal radiation of the materials made it possible to highlight two great classes of radiative behavior: that of electrically conductive materials (Copper), and that of dielectrics (granite and white plastic). The properties of the selected materials are presented in Table 2.

Table 2. A few properties of thermo-physical of selected materials

	mass density kg/m³	thermal conductivity W/(m.K)	specific heat J/(Kg)
Granite	2650	3,0~3,4	834
Copper	8920	401	385
White plastic	1 190	0.7~1,4	1050
Tedlar	1390	0.033	1400

In Fig. 2, we plotted the variation of the emissivity of the selected materials as a function of the wavelength (5-15 μm) for two different temperatures of the emitter; in blue T_e = 286 K and in green T_e = 310 K.

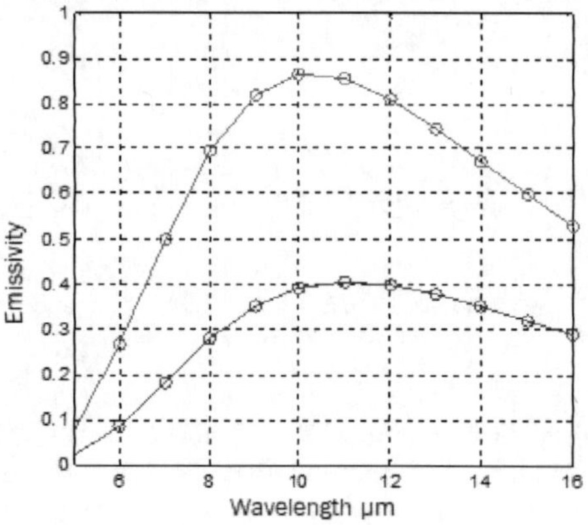

(a)

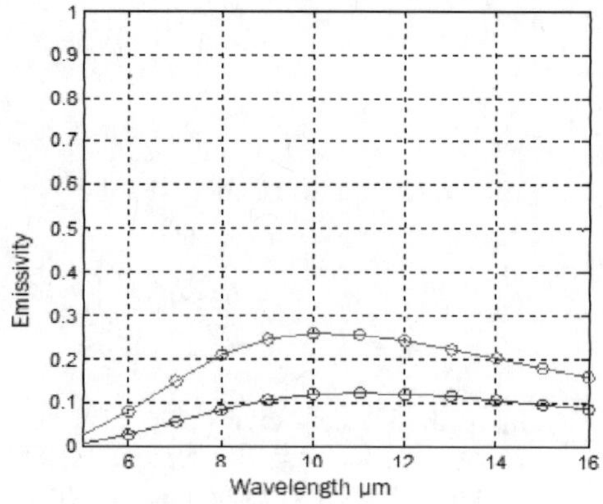

(b)

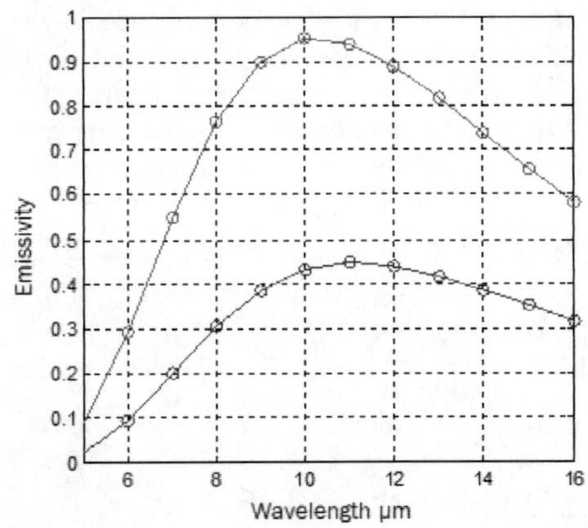

(c)

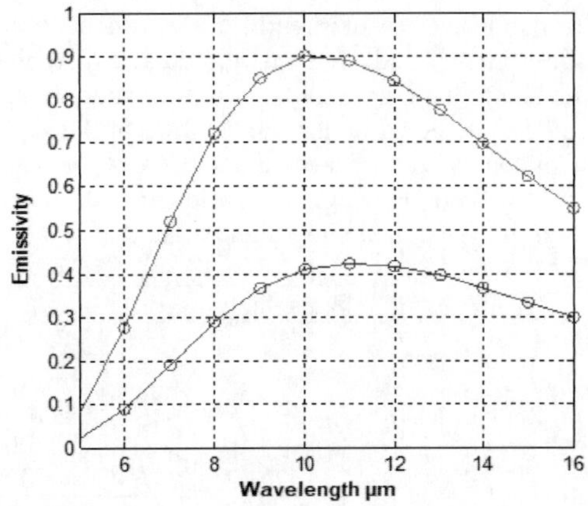

(d)

Figure 2. (a), (b), (c) and (d) are respectively the variation of the emissivity as a function of the wavelength of white, copper, granite and Tedlar plastics. In a clear sky and an atmospheric temperature equal to 285 K

The Graphs tell us, with a flat surface its temperature is equal to the atmospheric temperature, the emissivity does not exceed its initial value even if in a field where the rays can easily switch to the atmosphere and it is seen in all the curves in blue. This is correct, because the cooling by night radiation based on the principle of heat loss by radiation to the atmosphere until reaching a thermal equilibrium ($T_{at} = T_e$). In general, the monochromatic emissivity increases with the growth of T and λ in the atmospheric window and helpless when the atmospheric window when λ exceeds 13 µm. In order to compare the total monochromatic emissivity of these materials in the atmospheric window (8-14 µm), we have considered the following limiting case: $\tau = 1$ and $E_{at}(\lambda) = 0$ between 8-13 µm. where E_{at} is the atmospheric illumination and τ is the transmission coefficient.

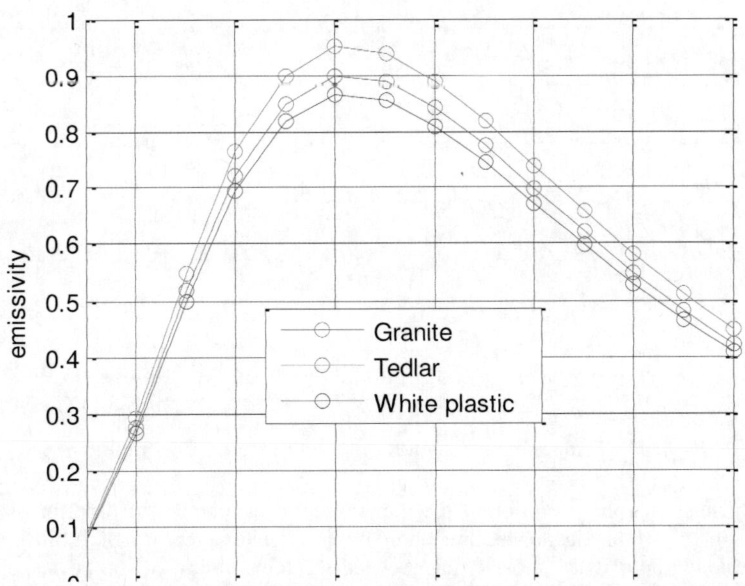

Figure 3. Comparison of the emissivity variation of selected materials between 8-14 µ.

From the results of Figure 2, we will choose the materials that have a great performance in the atmospheric window "Granite, Tedlar and White Plastic" in order. So to compare the variation of the monochromatic emissivity of these materials in the range of 5-18 µm with a transmitter temperature equal to 310K while the temperature of the sky is 285K. When the temperature of the emitter is very high compared to that of the atmosphere, the emissivity of the materials can reach its maximum values of 9 to 10 µm. However, it considerably decreases when this value is exceeded (Tab.2). The emissivity of granite is then slightly higher than that of Tedlar and white plastic. (Fig.3).

Wave length (µm)	Emissivity			
	Granite	Cuivre	White Plastic	Tedlar
8	0.76	0.20	0.69	0.72
10	0.95	0.24	0.86	0.89
12	0.88	0.20	0.80	0.84

Table 2. Emissivity values of materials in certain wavelengths

4. The temperature evaluation

In this section we assess the temperature of the selected materials as a function of time in the interval of 8-14 µ (Fig.3) and to simplify the calculation we have neglected the heat transfer by conduction as well the Effects panel. In the application of the variation of the internal energy of a plate characterize by; 0.06 m in thickness and 1m² surface area, we obtained the differential equation:

$$\frac{dT}{dt} + AT^4 + BT = C \qquad (10)$$

Where :

$$A = \frac{\sigma(\varepsilon_{sky} + \varepsilon_e)}{e \times \rho \times C_p} \ ; \quad B = \frac{h}{e \times \rho \times C_p} \ ;; \ C = \frac{hT_{amb} + \sigma(\varepsilon_{sky} + \varepsilon_e)}{e \times \rho \times C_p}$$

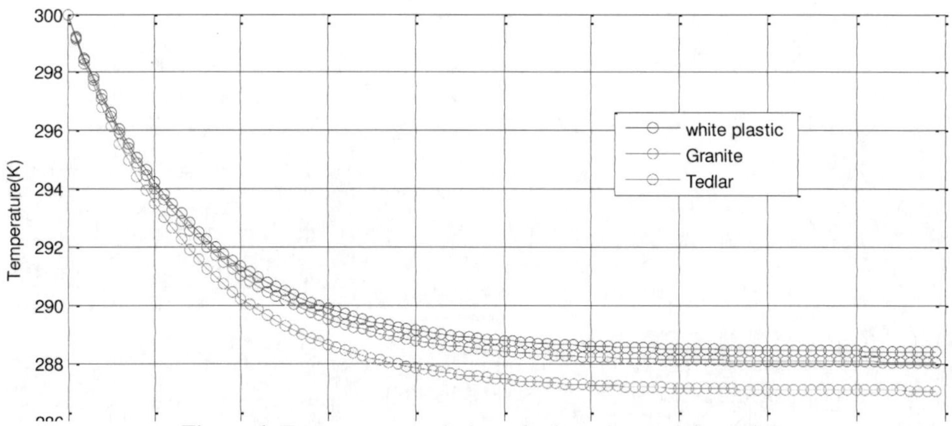

Figure 4. Temperature variation of selected materials at night

In figure. 4, we have shown the evolution of the temperature as a function of time for plates formed respectively by copper, white plastic, granite and Tedlar. The temperature of the different plates decreases with time. Granite is the most efficient since the temperature of the corresponding plate decreases by 13 ° C, this is due to its emissivity coefficient which can reach 0.95 in the atmospheric window. The temperatures of the Tedlar and white plastic plates decreased by 12 ° C and 11.7 ° C, respectively, for two minutes before stagnating. The temperature drop of the copper plate is 3.9 ° C. It is obvious then that the heat loss due to the considerable temperature difference in the night, will be able to provide cooling power in certain climatic zones in Morocco. This power will be optimized by a suitable choice of materials depending on the surrounding conditions.

5. Evaluation of the power of nocturnal cooling

We used the energy balance (Eq.3) to establish the expression of the net cooling power and to simplify the calculations we assumed that:

- The radiator is placed so that the conduction due to the ground is actually a convection ;
- The flow is zero because of the absence of solar rays ;

So the net cooling expression becomes on the following form:

$$\phi_{refr,nette} = \varepsilon_e(T_e)\sigma T_e^4 - \varepsilon_{ciel}\alpha_e(T_e)\sigma T_{ciel}^4 + h_c(T_{amb} - T_e) \quad (11)$$

The emission and absorption coefficients are given taken here as the average for all surfaces of the transmitter. In the choice of correlations to calculate the emissivity of the sky we considered that the state of the sky is clear, so we can calculate its value by:

$$\varepsilon_{sky} = 0.77 + 0.0038T_{dp} \quad (12)$$

Where T_{dp} and T_{sky} are respectively the dew-point temperature and the temperature of the sky. We can express this last by the following correlation:

$$T_{skyl} = 0.0522T_{amb}^{1.5} \quad (13)$$

And h_c is the coefficient of exchange we can express it by the following correlation [10] :

$$h_c = 2,8 + 3,3\,V \quad (14)$$

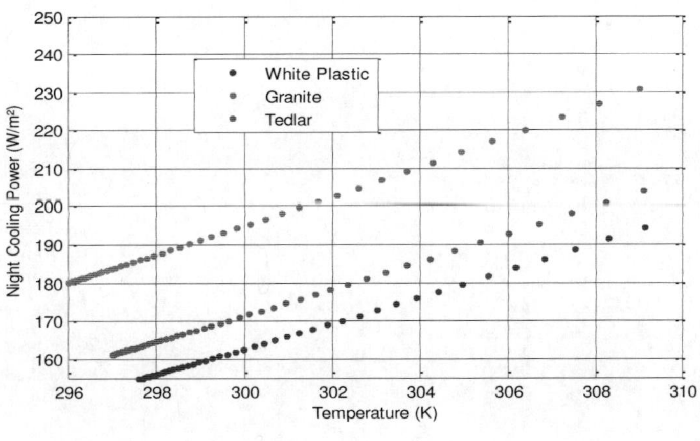

Figure 5. The variation in power as a function of transmitter temperature, in T_{amb} = 285K.

The decrease of temperature T_e is accompanied by a decrease in the cooling power, i.e. a drop of 13°C for granite, 12 °C and 11.7 °C for white plastic and Teldar respectively. As a result, the powers vary in a decreasing way up to a well-defined value of 180 W / m², 161.2 W / m² and 160 W / m², respectively presented by granite, Teldar and white plastic with an average difference of 19.4W / m² produced by the granite vis-à-vis compared to the other materials. These values reflect the equilibrium reached between the temperature T_e and T_{amb}.

References

[1] Energy Savings Potential of Radiative Cooling Technologies. (2015), (November).
[2] Sloant, R., Shaw, J. H., & Williams, D. (1956). Rsx-DsxlKnx,(1), 46(7), 2–6.

[3] Skies, F. C. (1981). A Set of Equations for Full Spectrum and a oTon above studies, 17(2), 295–304.
[4] Resources, W. (1975). On a Derivable Formula for Long-Wave Radiation From Clear Skies FLD), 11(5), 742–744.
[5] Begger, X., & Buriot, D. (1984). About the Equivalent Radiative Temperature, 32(6), 725–733.
[6] Sugita, M., & Surface, L. (1993). Cloud Effect in the Estimation of Instantaneous Downward Longwave Radiation, 29(3)
[7] Lhomme, J. P., Vacher, J. J., & Rocheteau, A. (2007). Estimating downward long-wave radiation on the Andean Altiplano, 145, 139–148. https://doi.org/10.1016/j.agrformet.2007.04.007
[8] Maykut, 1973. Report of the Joint U.S. POLEX Panel to the U.S. Committee for the Global Atmosphere [i.e. Atmospheric] Research Program (GARP), Committee on Polar Research (CPR), Committee on Atmospheric Sciences (CAS), Ocean Science Committee (OSC). USA
[9] Cuomo, V., & Silvestrini, V. (1975). The Radiative Cooling of Selective Surfaces, 17.
[10] Maamar Hamdani, Sidi Mohammed El Amine Bekkouche, Tayeb Benouaz, Rafik Belarbi, Mohamed Kamel cherier,'Minimization of indoor temperature and total solar insolation by Optimizing the building orientation in hot climate', Engineering structures and Techologies, N°6, pp 131-149,2014

IOP Conf. Series: Materials Science and Engineering **353** (2018) 012008 doi:10.1088/1757-899X/353/1/012008

Simulation of energy- efficient building prototype using different insulating materials

Salma Ouhaibi [1*], Naoual Belouaggadia [1], Rachid Lbibb[2] and Mohammed Ezzine[3]

[1] Laboratory of Engineering and Materials (LIMAT), Faculty of Sciences Ben M'Sik, Hassan II University of Casablanca, B.P 7955 Sidi Othmane, Casablanca, Morocco.
[2] Laboratory of Sustainable Development, Faculty of Sciences and Technics, University of Sultan Moulay Slimane, Beni Mellal, B.P 523 Beni Mellal, Morocco.
[3] Laboratory of Physico-chemistry of Applied Materials (LPCMA), Faculty of Sciences Ben M'Sik, Hassan II University of Casablanca, B.P 7955 Sidi Othmane, Casablanca, Morocco.

*E-mail: ouhaibisalma@gmail.com

Abstract. The objective of this work is to analyze the energetic efficiency of an individual building including an area of 130 m^2 multi-zone, located in the region of FEZ which is characterized by a very hot and dry climate in summer and a quite cold one in winter, by incorporating insulating materials. This study was performed using TRNSYS simulation software during a typical year of the FEZ region .Our simulation consists in developing a comparative study of two types of polystyrene and silica-aerogel insulation materials, in order to determine the best thermal performance. The results show that the thermal insulation of the building envelope is among the most effective solutions that give a significant reduction in energy requirements. Similarly, the use of silica-aerogels gives a good thermal performance, and therefore a good energy gain.

1. Introduction

The Moroccan Government aims to achieve a primary energy saving of about 12% on the 2020 horizon in the various economic sectors. Indeed, the building sector is one of the most energy-consuming in Morocco and rises up to 35% of the total power consumed by the country. This makes this sector a priority for the application of energy savings. The thermal insulation plays a main role in the reduction of energy consumption and also in the creation of a healthy and more comfortable space of life. That is why, a lot of research and studies are developed in order to improve the thermal insulation of buildings.

Among these studies, M. S. Mohsen et al. [1] focused on energy savings by building insulation using different materials such as polystyrene, rock wool and air gap. The results obtained showed that the use of polystyrene in wall and roof insulation could reach 77% energy savings, whereas rock wool cannot exceed 73%.

M. Ozel et al [2] studied the influence of the solar absorption of the external surface facing south, on the annual transmission loads of cooling and heating. The solar absorption coefficient of the outer surface has been varied from 0 to 1, with increments of 0.2, and extruded polystyrene was selected as insulating material. The results show that for insulated and non-insulated walls, the solar absorption coefficient influences very little the optimum thickness of the insulator, but has a greater effect on energy savings.

M. Ibrahim et al [3] has experimentally and analytically studied the thermal performance of exterior walls protected by an insulation layer based on «silica- aerogels ". The objective is to optimize the composition of the outer wall layers for the following cases: continuous heating, intermittent heating and no heating. For this purpose, six wall configurations were examined: insulation on the outside, insulation on the inside, insulation on the inside and outside, insulation in the middle, insulation in the middle and on the outside, insulation in the middle and on the inside. The results showed that the coating gives a better result than the traditional insulators.
Diagnosing the energy performance of a building provides information on the energy performance of a dwelling or building by evaluating the amount of energy actually consumed or estimated and its impact on greenhouse gas emissions. It describes the building or housing and its equipment for heating, air conditioning, for the production of sanitary hot water, cooling, ventilation, as well as the conditions of their use related to thermal comfort. It indicates, as the case may be, the amount of energy actually consumed (on the basis of the invoices), or the estimated energy consumption for a standardized use of the building or housing (with a method of calculation).
This is done initially by calculating the annual heating and cooling requirements per square-meter of the tested building, depending on the thermo-physical properties of its enclosure, the climatic zone, the rate of the glazed bays of the heated and/ or cooled spaces as well as their distribution on the various directions.
The chosen example is the diagnosis of energy efficiency (using the simulation software TRNSYS [4]), of an individual building having an area of 130 m^2. The local is located in the region of Fez. Insulating materials are tested in order to determine their impact on the energy consumption.
In this simulation, we examined the improvement brought by insulation of the roofs and the exterior walls by two types of polystyrene insulation materials and a silica-aerogels-based coating.

2. Description of the building

The project considers an individual house with an area of 130 m^2 including a bedroom, two living rooms, a kitchen, a restroom, a bathroom and a hall. It is located in the region of Fez, which is characterized by a very hot and dry climate in summer and a quite cold one in winter.
The composition of the walls constituting the building is described in the Table-1:

Figure 1. Scheme of the tested building

Table 1. Composition of the walls

Wall	Constitution (inside to outside)	U (W/m².K)
Outside wall	Coating (1,2 cm) + Brick (15 cm) + air layer (5 cm) + Brick (15 cm) + Coating (1.2 cm)	0,525
Roof	Mortar (2 cm) + Slabs 05_4 (15 cm) + Slabs 12_ Concrete (10 cm) + Mortar (2 cm)	0,813
Floor	Tile (3 cm) + Mortar (3 cm) + Concrete _c600 (20 cm)	0,893
Window	Double Glazing	2,95

3. Modeling

The dynamic behavior of the building is simulated through a multi-zone transient modeling by TRNSYS with a time step of 1h. We have divided the building into seven thermal zones. The simulation consists of calculating the cooling and heating requirements according to the standards applied in Morocco, the temperature of thermal comfort for heating is 20 °C and for air conditioning of 26 °C. The climatic data of the Fez region, taken from the METEONORM [7] software, have been integrated into the TMY2 form.
The figure below presents the modeling of the building studied in the software "TRNSYS simulation Studio".

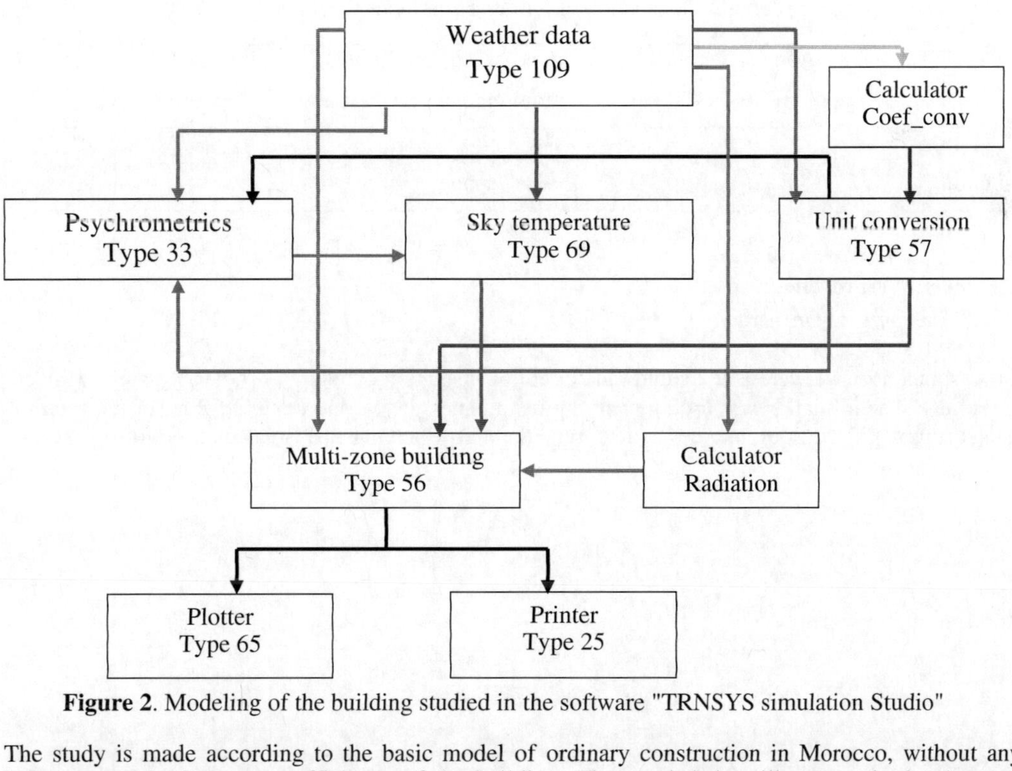

Figure 2. Modeling of the building studied in the software "TRNSYS simulation Studio"

The study is made according to the basic model of ordinary construction in Morocco, without any improvement in the energy efficiency of the building. The graph below illustrates the heating and cooling requirements of the building.

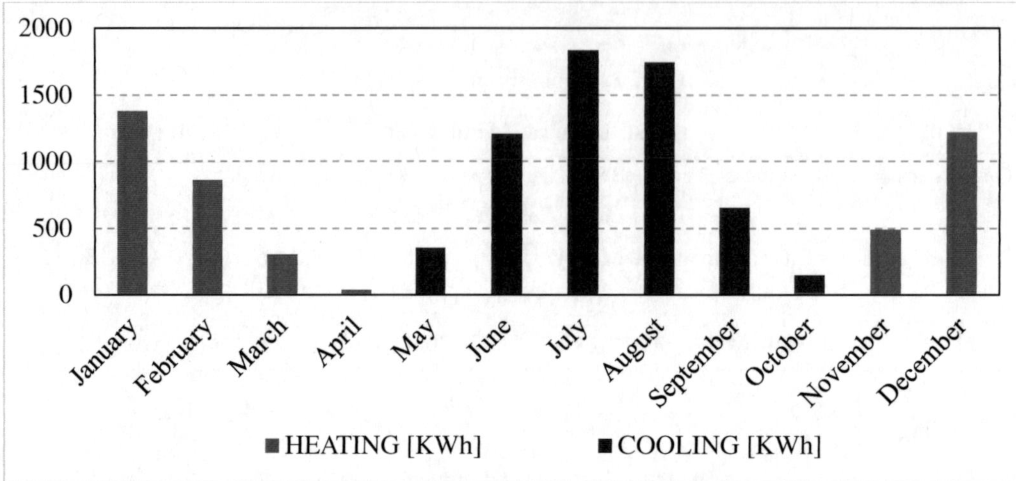

Figure 3. Monthly evolution of the building energy requirements

It is noted in Figure 3 that the heating period extends from November to March, while that of refreshment starts in May and ends in October. It may also be observed that the heating and cooling peaks correspond respectively to the months January and July.
By calculating the annual energy requirements, the results mentioned in Table 2, are obtained.

Table 2. Calculation of building energy requirements

	Annual Heating requirements (KWh/year)	Annual Air Conditioning requirements (KWh/year)	Energy Performance (KWh/m^2.year)
Tested building	4299	5912	78,54

4. Proposed improvements

4.1 The impact of insulation

In this simulation, we were able to study the effect of insulation on energy efficiency using extruded polystyrene. The insulation was used by varying its thickness between 1 cm and 6 cm, for the exterior walls, the roof and the floor, in order to determine the part which must be isolated as a priority.

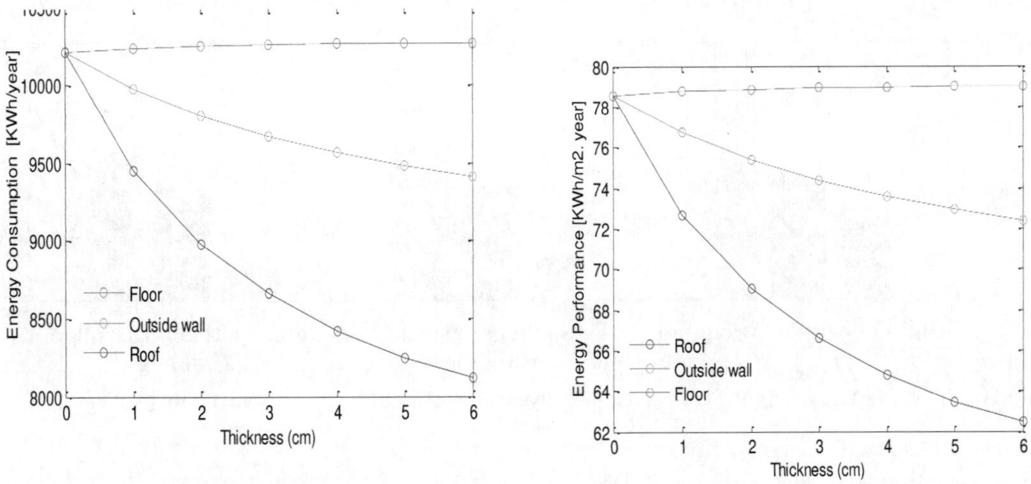

Figure 4. Energy consumption and performance according to the thickness of the insulator

In view of these results, it can be observed that the insulation of the roof and the outside walls is accompanied by a considerable decrease of energy requirement in heating and air conditioning when the thickness of the insulation increases. This need seems to stabilize from a thickness of 6 cm insulation. The insulation of the outside walls can thus bring an energy gain in the order of 7.80%, and that of the roof, contributes to a gain of 20.45%, thus much more important than the external walls. Wheras, floor insulation has no influence on energy consumption and performance. Since there is non-heat exchange.

4.2 Comparative Study: choice of insulating materials

Then the simulation is centered on the suitability of the choice of insulating materials. Thus, we choose for the outside walls and the roof two types of insulation: the polystyrene and a silica-aerogel-based coating,

Silica-aerogel are lightweight and have exceptional thermal properties (high porosity, low density and low thermal conductivity). The technique used is based on the application of such an insulating and innovative coating based on the aerogels of silica for the outside surface of the building in order to make a thermal insulating coating.

The characteristics of the materials are described in the following table:

Table 3. Characteristics of the insulating materials used

Materials	Thickness (cm)	Thermal conductivity (W/m.K)	Density (Kg/m3)	Specific heat (J/Kg.K)
Polystyrene	6	0,030	35	1180
Silica-aerogel-based coating	6	0,027	200	1100

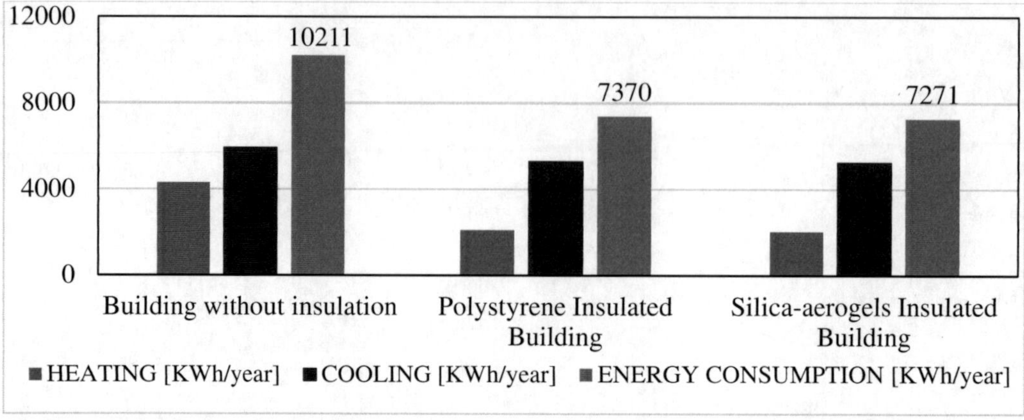

Figure 5. Calculation of the building's annual energy requirements according to the insulation material

The results of the calculation show a clear improvement in terms of energy requirements when using innovative insulating materials, since energy gain can reach 29% using silica-aerogel-based coating and 28% using polystyrene

4.3 Effect of the external surface temperature on the transmission loads

<u>Mathematical model</u>

Considering an external wall subjected to several variations of parameters, ambient temperature, solar radiation, convection and radiation, as shown in the diagram below:

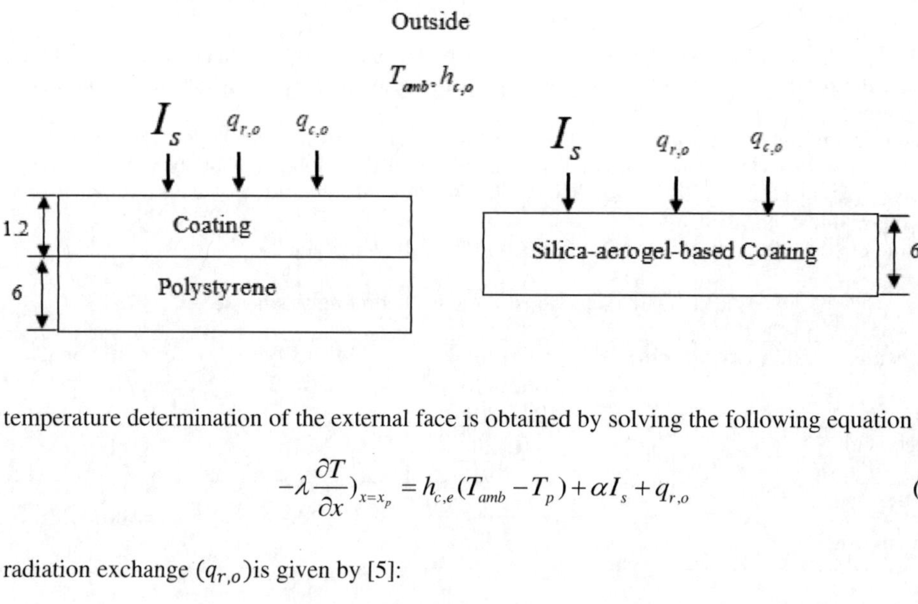

The temperature determination of the external face is obtained by solving the following equation [5]:

$$-\lambda \frac{\partial T}{\partial x})_{x=x_p} = h_{c,e}(T_{amb} - T_p) + \alpha I_s + q_{r,o} \quad (1)$$

The radiation exchange ($q_{r,o}$)is given by [5]:

$$q_{r,o} = \varepsilon_0 \sigma (T_{sky}^4 - T_p^4) \quad (2)$$

Where ε_0 is the surface emissivity appropriate to materials, and the sky temperature can be calculated as follows [6].

IOP Conf. Series: Materials Science and Engineering **353** (2018) 012008 doi:10.1088/1757-899X/353/1/012008

$$T_{sky} = 0,0552T_{amb}^{1,5} \tag{3}$$

The convective heat transfer coefficient has been estimated according to [6] :

$$h_{c,o} = 2,8 + 3,3V \tag{4}$$

Where:

$q_{r,o}$:Radiation density of heat flux;

$h_{c,o}$:Convection thermal transmission coefficient ;

V:Wind speed;

ε_o:*Emissivity of the material;*

σ: *Setefan-Boltzmann constant;*

α:*Absorption* coefficient;

T_{amb}:Ambient temperature;

T_{sky}: Sky temperature;

T_p:Outside surface temperature;

Discrete equation:

$$-\lambda \frac{(T_i - T_{i-1})}{\Delta x} = h_{c,o}(T_{amb} - T_p) + \alpha I_s + q_{r,o} \tag{5}$$

With:

$T_i = T_p$

$T_{i-1} = T_{amb}$

$\Delta x =$ The mesh step

We can also determine the outside surface temperature of the roof:

$$\alpha I_s + \varepsilon_0 \sigma (T_{sky}^4 - T_p^4) + h_{c,o}(T_{amb} - T_p) + \frac{\lambda}{\Delta x}(T_p - T_{amb}) = 0 \tag{6}$$

For solving this equation, we use the Newton-Raphson method that is generally used in non-linear one-dimensional problems. Thus, the temperature profiles of the external face are represented in Figure 6.

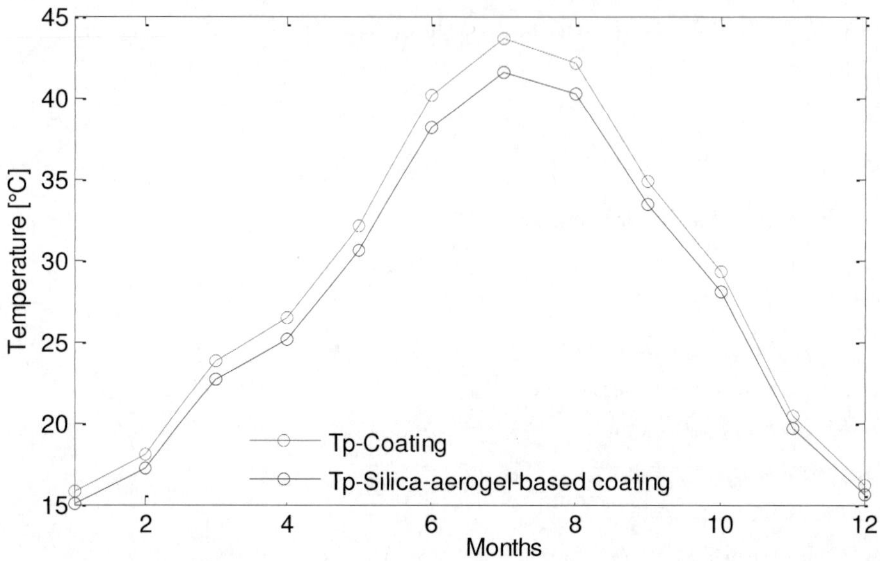

Figure 6. Monthly Evolution of the outside surface temperature

It can be observed that the outside surface temperature follows the same trend as in the case of a coating with the silica-aerogel-based material. However, this temperature remains slightly higher in the case of coating (with a relative difference of 5%). This is probably due to the large amount of heat transformed from the absorbed solar radiation producing a temperature rise of the outside surface, therefore the internal temperature will also be relatively higher. We can therefore conclude that the outside surface temperature has a significant effect on the annual heating and cooling transmission loads.

5. Conclusion

This work has been devoted to the study of the conformity of the energy performance of a multi-zone building located in the region of Fez.
The heating and cooling energy requirements of this building were calculated on the basis of the thermo-physical properties of the enclosure and their distribution in the various orientations.
The conclusions are:

- ✓ The insulation of the roof and the external walls leads to an energetic decrease in heating and air conditioning. Similarly, the insulation of the roof seems to be more important compared to the insulation of the external walls.
- ✓ The energy gain can reach 29% using the silica-aerogels base coating and 28% using polystyrene.
- ✓ The temperature of the outside face has a great effect on the annual heating and cooling transmission loads.
- ✓ In the event that the house is heated and air-conditioned by a stationary electric source produced from fossil fuel, the insulation of the roof and exterior walls leads to a reduction of green house gas emissions in the range of 1900 Kg CO2 eq per year. According to the same hypothesis, the generalization of this isolation alternative to all residences of the city of Fez would be a real opportunity to decrease energy consumption up to 200 000 T-CO2 eq per year. This type of initiative supports government action, enhances the efforts of the kingdom's territories in favor of the climate, and respects Morocco's commitments of reducing its greenhouse gas emissions by 42% on the 2030 horizon.

IOP Conf. Series: Materials Science and Engineering **353** (2018) 012008 doi:10.1088/1757-899X/353/1/012008

6. References

[1] M.S. Mohsen and B.A. Akash, '*Some Prospect of Energy Saving in Building'*, Energy Conversion Management, N°11, pp. 1307 - 1315, 2001.

[2] Meral Ozel,'The influence of exterior surface solar absorptivity on thermal characteristics And optimum insulation thickness', Renewable Energy, N° 39, pp. 347-355,2012.

[3] Mohamad Ibrahim, Pascal Henry Biwole, Etienne Wurtz, Patrick Achard, 'A study on the Thermal performance of exterior walls covered with a recently patented silica-aerogel-Based insulating coating', Building and Environment, N°88, pp. 112-122, 2014.

[4] TRNSYS version 16.09, "Transient system simulation program developed by solar Energy Laboratory", University of Wins consin Madison, USA TRANS solar Energietechnik gbbh and centre Scientifique due Batiment and Thermal Energy System specialists, LLC, 2002-2005.

[5] Sami A. Al-Sanea,'Finite-volume thermal analysis of building roofs under two- Dimensional periodic conditions', Building and Environment, N°38, pp 1039 – 1049,2003.

[6] Maamar Hamdani, Sidi Mohammed El Amine Bekkouche, Tayeb Benouaz, Rafik Belarbi, Mohamed Kamel cherier,'Minimization of indoor temperature and total solar insolation by Optimizing the building orientation in hot climate', Engineering structures and Techologies, N°6, pp 131-149,2014.

[7] METEONORM 7.1 (2014) Software application available from Meteo test, Fabrikstr 14,3012 Bern Switzerland (www.meteonorm.com).

IOP Conf. Series: Materials Science and Engineering **353** (2018) 012009 doi:10.1088/1757-899X/353/1/012009

Desert rose: building material of cupolas in the Souf in Algeria

C AZIL[1], B DJEBRI[1] and L ROVERO[2]

[1] Laboratorie "Ville, architecture et patrimoine", "Ecole polytechnique d'architecture et d'urbanisme d'Alger EPAU", Algeria;
[2] Department of Architecture, Materials and Structures Division, University of Florence, Italy.

Email: c.azil@epau-alger.edu.dz

Abstract. In the Souf of Algeria, the roofs of all constructions are arranged like corbelled domes, built with local particular material to this region, which is the desert rose. These cupolas describe a unique landscape of historic centres. Such constructions include a widespread and precious heritage that deserves protection to save this urban landscape which constitutes an element of identity of heritage built upon the material as well as the immaterial of the local know-how. Unfortunately, these architectural elements have undergone alterations that devalue the urban landscape and destabilize the buildings. However, the structural system that provides stability and endurance to this day remains an open question.

In this, paper, we describe the role of desert rose cupolas in the construction of a single urban landscape and we contribute to this knowledge. Then, we explain the role of the availability of the materials locals (desert rose and tafza) to appearance ad emergence of construction with cupolas typology. In addition, we describe these materials locals, and the method to them usage. In the end, we have traced the process of construction of these cupolas by corbelling which is mounted by successive courses of the desert rose and the plaster mortar.

Keywords. Dome, Desert rose, Tafza, structural analysis, local material, building heritage.

1. Introduction

In the Mediterranean basin, several cities are known by the use of cupolas in all constructions. The latter constitute true heritage values to be protected, and to transmit this know-how to future generations.

In the Souf region of Algeria, all buildings has cupolas. This use has appeared since the founding of three historical core in this region namely *Guemar*, *Z'goum* and *el Oued* towards the end of the XVI century [1] .The cupolas of this era are built with desert rose material.

These local materials played a large role in the construction of a particular vernacular architecture specific to this region, especially through the use of domes.

Consequently, these major architectural elements describe the urban landscape of this region, which is an identity element of the culture in the Souf region. The damage suffered in all these cupolas devalues this urban landscape.

Therefore, it is important to recognize this material its sources, and the way of its use for the construction of domes and in some cases the walls of these vernacular dwellings.

IOP Conf. Series: Materials Science and Engineering **353** (2018) 012009 doi:10.1088/1757-899X/353/1/012009

2. Materials locals: a factors of emergence of architectural typology " *cupolas*"

The old-style landscape of the Souf is denoted by a particular beauty due to the originality of the site "a human establishment produced in an erg". The splendor does not count only in the movement of the ocher or white dunes of the erg, but also in unexpected human creation [1].

The specificities of a built environment are severely related to the characteristics of the human and natural environment to which they relate, and some of their architectural and constructive features can be found in the conscious choices made by builders in relation to the context of and the presence of exceptional natural phenomena. Therefore, we expose the natural factor that contributed increase to this typology particular to the Souf region in the use of domes.

2.1. Factors of emergences: Form, technique and environment

The architecture of the desert in Algeria is vernacular. The construction of the Souf was carried out without the technical support of an architect or an engineer.

Vernacular architecture was considered by Giancarlo Cataldi as typically the fruit of the indissoluble union between a given natural environment (context) and a determined human culture [2]. We interest at the first factor which is natural factor.

By crossing the correlation between technique and form, he considers building materials as a major factor influencing the development of an architectural and structural typology [2]. As a result, in the Souf region, the lack of wood, which allows large crossings, has eliminated the use of flat roofs in this region. On the other hand, the presence of a material, which has a certain adhesion and a resistance to compression important by its architectural morphology, which is "the desert rose ", which determined the constructive typology of the relative cupolas [3].

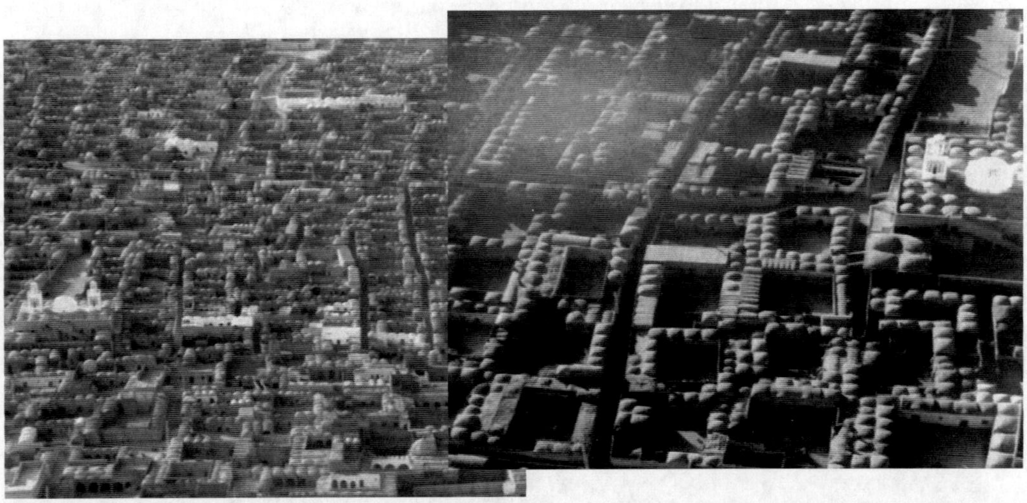

Figure 1&2. Old quarter in Guemar, Souf [4]

Over time, the crossing of these multiple factors, namely: the function of a structure, the static system, building materials and the historical era, gives rise to the consolidation of architectural forms and technical solutions characterizing precisely this environment, in symbiosis with this human society that is established there.

3. Building Materials local : Sources and Methods

3.1. Building materials: provenance

The unusual architectural sense of the *Souf* is due principally to the variety of the particular materials existing on the spot. The ground and the subsoil of the Souf have impressive opposites, there are frequent sand in the erg, *tafza_* limestone _, and the desert rose _ or *Lous_* [5]. These two building materials are scattered in the soil and the subsoil.

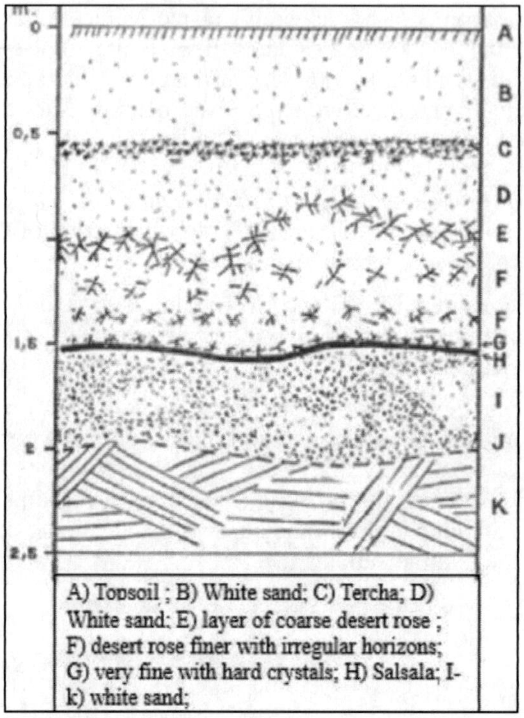

Figure 3. Schematic soil section [6]

In addition, the gypsum solidification of the region produces rocks of various sizes and shapes, namely, the desert rose - *Lous-* . it is made of interleaved spearhead crystals up to 20 to 30 cm in length, solid building stone in the form of coarse rubble very resistant to building [7]; *Tafza*, which is a white sandstone containing very little sand, often quite hard. The latter constitutes the raw material of plaster, it is used as binder and coating [6]. We quote that there are other materials used in other structural elements in the image of the *Salsala* used in for foundations and walls.

Figure 4. The desert rose material [5]

Figure 5. The material of tafza [5]

3.2. Building materials: preparation

Each material has its own way of preparation so that it can be used during construction. The desert rose apparent and flush with the ground is picked up by hand. However, extraction work is necessary

IOP Conf. Series: Materials Science and Engineering **353** (2018) 012009 doi:10.1088/1757-899X/353/1/012009

if the stone is deep; several tools are used for this work, such as the dynamite process, due to the hardness of this material often used because of its strength and its impermeability in these construction types (dome) [5].

By passing to the material *tafza* which serves to produce the local plaster. It is necessary to rid all the upper layers that can be constituted of the desert rose or of the sands to arrive at the compact mass of *tafza* in the ground. The deep layers of the latter, which are in contact with the water table, are moist, so it needs to be dried once extracted; on the contrary, the upper layers are dry. The steps of extracting it begin with the disengagement of large blocks of the ground, by performing cuttings to obtain parallelepiped kinds. Once isolated this mass is broken into stones of smaller dimensions in order to go to the oven for cooking. After two hours of cooking *tafza*, it takes at least two days for the finished product to cool down. *Tafza* crumbles inside the oven but this work ends by reducing it to powder [5].

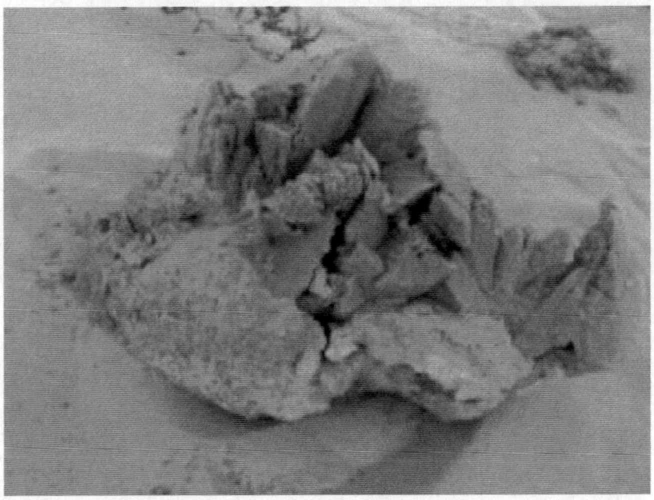

Figure 6. Stone and mortar of a wall of an old rural house in the Souf [5].

3.3. *Methods of building cupolas of desert rose*

By the processes linked to the previous materials, the cupolas were made easier to build. It is the unit of design for the interior space (same unit of size of the rooms and the roof). Juxtaposed cupolas can be multiplied to increase the number of adjoining rooms (twice or more) and enlarge the spans [5].

Figure 7. House of the Souf area [1].

It is self-supporting cupolas that are slightly flattened (lowered cupolas), or its height is about half its diameter. It is constructed by corbelling, mounted without formwork by successive courses [7]. It

IOP Conf. Series: Materials Science and Engineering **353** (2018) 012009 doi:10.1088/1757-899X/353/1/012009

cannot exceed a diameter of 2.50m. Therefore, the gauge of the parts is closely related to the roof gauge [1]. The shell begins with a thickness of 20 cm and decreases slightly at the top [5].

For its realization, using the material desert rose and plaster, the mason ascends the successive courses, in corbelling one above the other, until it ends in the center. This conduct of the construction is made possible because this plaster has a very fast grip [1]. Its shape of basis is square. The execution of this copula is made on a base composed by walls or arches flattened at the same level. Then, by building small squinch with plaster and handle by handle. After, sitting up sitting with a material of a rose of sands and plaster. This cupola is much flattened, or this form is close to a cloister vault, but its edges would be curved. By coating these cupolas with an exterior coating. Some cupolas are not coated.

Figure 8. A cupola of a rural house [8].

Figure 9. A cupola of a rural house [5].

The base of the dome is sealed, in these quarters side by side with the walls of the space concerned which in one sense, can have an arcade interruption supporting one side of the dome. The stones at the top of the walls allow the hanging of the copula [5].

In the Mediterranean region, architecture dome, building with local materials, has an important place. Among these architecture, it is important to mention Syrian architecture. The latter is built with the local material "the earth". The cupolas of the architecture of Syria are built without formwork, with the corbelled method. The shape of these cupolas is conical, it is different in the shape of the desert rose domes. The factors of emergence of the dome architecture in Syria are similar to those of the architecture of the Souf, which are because of the absence of wood [9].

Figure 10. A dome of Syria region [9].

3.4. Other use of building material local: the desert rose for building walls and foundations

The Desert rose is also used in the building of the other elements in the *Soufi*'s houses.

Figure 11. An old rural house building by desert rose [5].

3.4.1. Foundations. This stone is used in building of foundation of houses. Since the Soufi house was often a one-storey house, the foundations are almost superficial and of course continuous, with a thicker width (45 to 50 cm) than that of the wall (30 to 35 cm) with a depth of from 50 to 70 cm. This foundation made of desert rose stones and plaster. The desert rose stones used in the foundations are hard and large.

Figure 12. Foundation of old house with desert rose [5].

3.4.2. Walls. The desert rose is also used in the construction of certain walls bearing the rural houses of the Souf. The walls of houses are usually 0.30 m thick, a height of floor between 2 to 2.5 meters.

Figure 13&14. Profiles of walls of old house with desert rose [5].

The stones is bound with a plaster mortar. The shape of the stone is very irregular, so it needed to fill all the interstices with plaster and smalls stones.

Figure 15. Texture of walls of old house with desert rose [5].

It clear that such material has added important values in the architecture and construction of the desert in Algeria, because thanks to him, builders can build houses with simple principles without the need for formwork, nor the need for polished. Thanks to its resistant the thicknesses of the walls and cupolas are not very big. Its texture in the construction adds an aesthetic value thanks to its form decorated by relief gathers to roses. On the other hand, the Souf's house is also known for its pleasant thermal comfort in the different seasons thanks to the use of copulas. This copula is a conceptual solution, which considered as an isothermal device because this form will heat up less quickly because the volume of the room with copula is large.

4. Conclusion

Through this first contribution to the knowledge of material local of the region of the Souf, namely: the desert rose and *Tafza* stone. The historical analysis gave us that availability of this locals materials

IOP Conf. Series: Materials Science and Engineering **353** (2018) 012009 doi:10.1088/1757-899X/353/1/012009

constitute factor of appearance of typology architectural, which is building cupolas. In addition, we have described these materials, their provenance, their preparation and how they are used. In the end, we have traced the process of construction of these cupolas by corbelling which is mounted by successive courses of the desert rose and the plaster mortar. Therefore, we explain the other elements of building, which is constructed by this the desert rose material, as like a foundations and walls. After this study about this material ad their role for construction of an important architectural vernacular heritage, it is necessary to study of the contribution of this material to hydrothermal comfort of the constructions.

References

[1] Coté M, 2006 Si le Souf m'était conté. Comment se fait et se défait un paysage, Média-Plus, Constantine, p 135.
[2] Cataldi G, 1989 Attualita del primitivo e del tradizioale i architectura. Aliea, Fireze,p 199.
[3] Ferry, 1951 Série moographies. Sahara Souf, Documents Algériens , Série Monographie, Alger, http://alger-roi.fr/Alger/documents_algeriens/monographies/pages/9_souf.htm, accesses on 12/01/2017
[4] Sebe A, 1989 Moula-Moula: Le Sahara à vol d'oiseau, tagoulmoust, p180
[5] Lebsir A, 2016 les cultures constructives traditionnelles, université mohamed khider , Biskra, p292
[6] Bataillon C, 1955 Le Souf. Étude de géographie humaine. Université d'Alger, Alger, p 190.
[7] Mangin, D, Panerai, P, 1999 Institutions et formes urbaines. Guémar et Tamelhat,Algérie.
[8] https://fr.wikipedia.org/wiki/Vo%C3%BBte_en_arc-de-clo%C3%AEtre , seen 03/14th/2018
[9] Mecca S, Dipasquale L, 2009 Earthen Domes et Habitats. Villages of Northern Syria, Edizioni ETS,

IOP Conf. Series: Materials Science and Engineering **353** (2018) 012010 doi:10.1088/1757-899X/353/1/012010

The Erosion of expansive soils through a zoning of Taza (Morocco)

A. DEMEHATI[1*], A. ABIDI[1], M. EL QANDIL

[1] University Sidi Mohamed Ben Abdellah, Faculty of Sciences Dhar El Mahraz, Department of Geology, P.O. Box 1796, Fez, Morocco

E-mail*: abdelghani.demehati@usmba.ac.ma

Abstract. Like other cities in the region of Fez-Meknes, Taza undergoes the phenomena of erosion of soils facing the builder in general and the town planner in particular. This type of aggression of the land is sometimes amplified by the swelling-withdrawal of soils. In effect the marl clay layers are considered as unstable soils because of their volumetric changes depending on the duration and intensity of extreme climate changes (weather, droughts, ventilation.). This phenomenon, which is part directly or indirectly so-called natural disasters, can lead to significant costly damage and particularly intense (soil erosion, landslides, obstruction of the course of the water, hydraulic works and sanitation networks, cracks of the constructions and their annexes', deformations of pavements, etc.). The knowledge and mastery of this phenomenon are essential to prevent, limit or remedy the consequences and damage related to their impacts. We focus as well our contribution on the geotechnical characteristics of the soils of the city of Taza with highlighting of the hazards in susceptibility to erosion of soils in particular through their potential of swelling-removing and this for the benefit geotechnical zoning of Taza.

1. Introduction

The Moroccan area of Fez-Meknes and for numerous cities as it is the case of Taza, the urban development coped with multiple challenges, because of the extensions in mass for the realization of the infrastructures, works, buildings and other installations. The new constraints of safeguard of the environment and the climate changes sharpen the intensity and the degree of the risks of the natural disasters type. The geological and geotechnical factors are determining on this subject and can generate catastrophic effects if they are badly taken into account or if the projects lack of adaptation to the sites. On the level of Taza, the parameters of swelling and withdrawal of the grounds predispose on the whole with these phenomena. Direct and indirect methods make it possible to consider as well risk of erosion as that related to swelling. Among the indirect methods, we have empirical procedures, which make it possible to exploit physical or chemical results carried out on representative selected samples. The direct methods are in several procedures of mechanical tests and are carried out on undisturbed samples. The œdometric test is that which is commonly used to measure the expansibility and the pressure of swelling. Other methods consider the loss due to erosion or the percentage of swelling starting from the chemical characteristics and mineralogical analyses. In this article we referred to the geotechnical current tests (Atterberg limits, swelling stress, value of Blue) [1], on the one hand and the cohesion of the ground on the other hand. Our objective is consequently to show that more the grounds are likely with the swelling-withdrawal more they are potentially erodible. After presentation of the site of the city of Taza, we recall various types of soil erosion that may affect the structures or lands concerned and various correlations proposed by different researchers. From the site we have selected marly samples to test them in the laboratory. We have then considered correlations between the Atterberg limits, the swelling pressure, the cohesion and the loss due to erosion.

2. Urban site

2.1. Taza city context

Taza city is located north-east of Morocco and east of Fez- Meknes area. It fits between Prerif lies to north and of the Middle Atlas to the south like structural fields. The pre-Riffian zone [2,3] constitutes the southernmost part of the external furrow of the chain of Rif. It is made of grounds of Tertiary age represented by a series of marl blue, very powerful, relatively homogeneous of higher Miocene (Figure 1) and the olistholites with varied ages and natures. The deposits of the Middle Atlas contrast completely with those of the Riffian pre-zone. They are composed of Liasic limestones and marl-limestones resting onargillaceous deposits of permo-triasic age, intersected with outpourings doleritic [5].

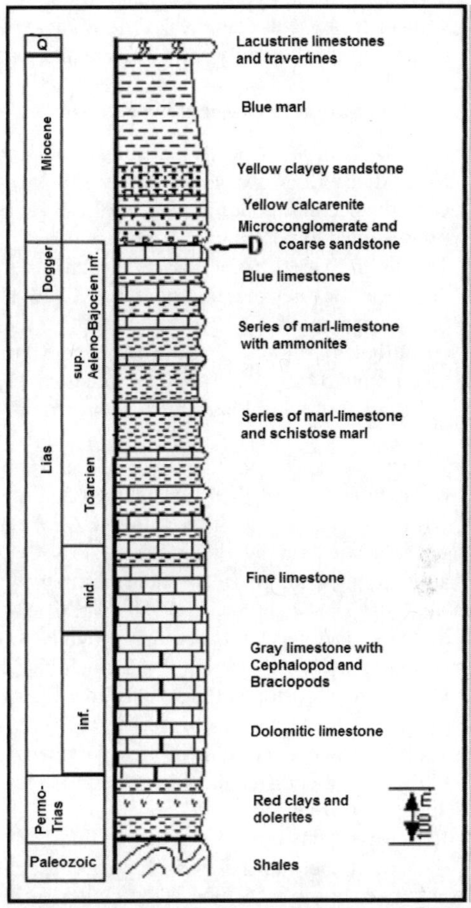

Figure 1. Synthetic log of Taza region [4].

2.2. Climate of Taza

Taza province climate is semi continental type with Mediterranean influence. It is marked by a temperate and humid winter season (October to May), and hot and dry summer one (June to September). Temperatures vary between -8° and 38°, while rainfall ranges between 100 and 700 mm / year [6].

These conditions put the grounds at severely tested of alternate of water content and consequently volumetric variations. It is what will be clarified in the continuation of our matter.

3. Types of erosion

Often types of erosion are rather undesirable that we consider individually.

3.1. Undermining of the hydraulic works

3.1.1. Channels. One of the estimates of the constraint of training of blocks of protection in a single channel with ripraps is given by [7]:

$$\tau = \left(\left(\frac{tan(\phi)}{0.03}\right) \Upsilon w \; tan(i)\right) \quad (1).$$

With ϕ : angle of friction of the ripraps.Υw: voluminal weight of water. h: height of water. i: slope of the flow.

3.1.2. Structure on waterway

The depth h of undermining is estimated by LPEE [7]:

$$h = 0.217\,(d50)^{2/7}\,(Q/L)^{6/7} \quad (2)$$

With Q=flow of the wadi. h= height undermining compared to the coast of high waters. L=width of the bed of wadi. d50=average diameter (sieve letting pass to sifting 50% of material).

3.1.3. Dam. For estimation, the speed of silting of the stoppings as result of erosion of the basins slopes in their upstream one can resort to the formula of Baker [7]:

$$D = 6.14C - 49.78 \text{ (t/km}^2\text{/an)} \quad (3).$$

Where C is the coefficient aggressiveness climatic: $C = P^2/Pm$. P: precipitation of the wettest month. Pm: annual precipitation averages.

We note that the hydromechanical effect of erosion is important as soon as the angle of friction of the soils or their mean diameter of the grains is low and the monthly rainfall is momentarily intense. This is verified for the marly soils of Taza and its region and their climate.

4. Correlation between characteristics

4.1. Correlations between parameters of identification and erosion

We initially refer on the correlations of Mr. Goutte-Lima [8]. In reference to this experimentation we can advance that under the effect of the action of erosion the loss of the water-logged soil is inversely proportional to cohesion.

Just as the variation of loss of ground is in quasi proportionality with the variation of the loss of cohesion. Besides the digraph of Mc Graw-Hill [7] makes it possible to envisage fields speeds leading to sedimentation, or on the contrary transport until the erosion of the ground following its granulometry highlighting the aggressive dynamic action of the flow on the haulage of the grounds.

According to Dos Santos and E. Decastro [7] they consider criteria based on the granulometry and the swelling of ground. Philipponat proposes the quantification of the impact of erosion by [7]:

$$Er = H + P \quad (4)$$

H (in cm): represents the depth of the drain dug in a test tube of ground subjected to thin fountain constant speed and flow during a given time. P (in g): weight of the moved ground. A. Merzouk [9] indicates that the role of the mineralogical characteristics on erosion of nine Moroccan soil types could be approximated by the index of loss of ground (Sl) for a slope of 9% and under a shower of simulation of rain of 10 cm/ hour of intensity:

$$Sl = 311.83 - 4.48(Sm + S) + 6.45C + 613.40Ec \quad (5).$$

With Sl=index of loss of the ground; Sm=matrix of the ground<2mm; S=%sand; Ec= electric conductivity of the ground. C=% of active CaCO3.

Considering erosion at the level of the fields the quantification is obtained thanks to the formula [7]:

$$A = K\ R\ L\ S\ Cp \quad \text{(Wischmeier and Smith)} \quad (6)$$

With A= erosion per unit of area ;K=coefficient erodibility of the ground ;R=coefficient influence of the rain; LS=coefficient of geometry of the watershed (length-slope) ; Cp = coefficient of vegetation and protective works.

It follows from this research that fine soils of the Taza marl type will be more sensitive to water erosion if their cohesion is low because they are less compact due to volumetric shrinkage, for example. In all cases, the steep slopes of this city and its region contribute to the predisposition to thrusting.

4.2.Correlations between parameters of identification and swelling

Komornik and David [10] propose a relation between the pressure swelling σg in kPa, the dry density γd, the water content W and the liquid limit WL:

$$lg(\sigma g) = 0.0208WL + 0.000665\gamma d - 0.0269W + 2.132 \quad (7).$$

The model of Brackley [11] proposed a general relation between the pressure swelling in kPa, the index of the vacuums e and the index of plasticity IP of the ground:

$$lg(\sigma g) = 5.3 - (147\frac{e}{IP}) \quad (8)$$

When M. Chrétien[12] drew up a relation between the index of plasticity IP and the specific value of blue VBS of a clay inflating according to a correlation of about 0.979:

$$IP = 10.51\ VBS + 6.948 \quad (9).$$

Of the same Chen [13], being based on work of Kassif and Baker [14], concluded that the pressure of swelling is not affected by the water content in the event of prevention of the variation of the dry voluminal weight.

In addition, work of M. Goutte-Lima and al. [8], arises a susceptibility of the wet argillaceous-muddy grounds saturated to erosion is all the more notable as cohesion is notable on the one hand, or that the loss of cohesion, by withdrawal for example, is also important. The correlation in these cases is very good that we improve by proposing a linear function (Figures 2,3). Thus, the increase in the water content of the marl reduces its swelling pressure in favor of the increase of its void index, a reduction of its compactness and its cohesion which prelude to a weakness of resistance to erosion. as we show by our following experiments and correlations.

IOP Conf. Series: Materials Science and Engineering **353** (2018) 012010 doi:10.1088/1757-899X/353/1/012010

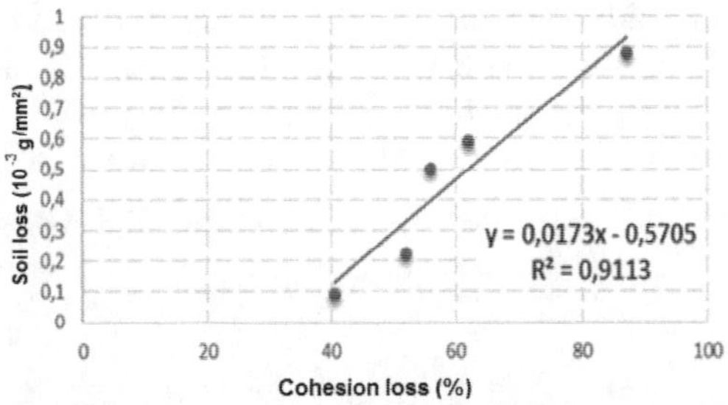

Figure 2: Decrease of the cohesion of the saturated clay soil related to the percentage loss by erosion according to Goutte-Lima and al. [8] modified

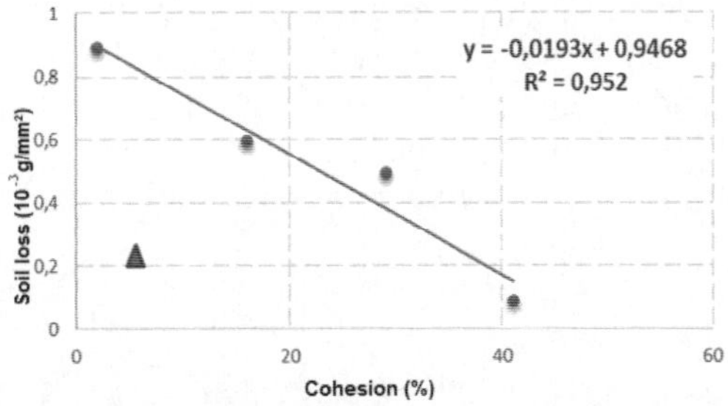

Figure 3: Evolution of the loss of ground by erosion according to the cohesion of the saturated clay soil according to Goutte-Lima and al. [8] modified

5. Fields of investigation and experimentation

In reference to the geology brought closer to the topography of Taza we carried out a set of samplings to marl predominance established and selected according to the transfer on the following map. Our samples cover thus mainly fields of slopes between 5 and 15% places of the most urbanized or potentially urbanizable, close to the road infrastructures, works and waterways, and thus threatened by the slips, the erosion of surface or undermining under effects of water streaming (Figure 4)

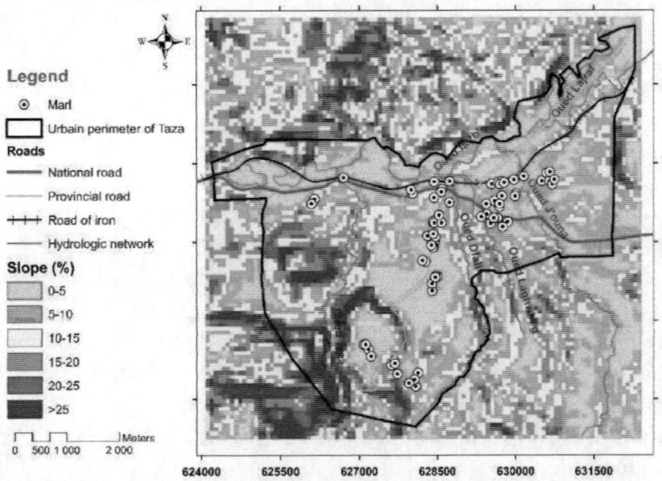

Figure 4: Location of experimental samples on slope map

6. Results and discussion

6.1. Geotechnical characteristics of the tested marls

Table 1 summarizes the minimum, maximum and medium values of the geotechnical characteristics of the one hundred marl samples tested: Atterberg limits WL and IP, mechanical characteristics not drained by shear cohesion (Cu), angle of friction (φu), as well as swelling pressure σg.

Table 1: Results of geotechnical characteristics of marls

	Blue marl			Green marl		
Parameters	MAX	MIN	MED	MAX	MIN	MED
WL (%)	69.00	15.47	47.22	70.00	29.00	49.93
IP (%)	44.00	17.00	28.02	48.00	13.00	28.87
Cu (kpa)	430.00	8.00	250.00	370.00	80.00	182.75
ϕu (°)	34.00	10.00	20.86	33.00	8.00	19.13
σg (kpa)	350.00	0.00	48.69	500.00	6.30	124.07

6.2. Comparison between plasticity and the value of Methylene blue

By referring to figure 5 graph, the results of marl and muddy samples of Taza related to the index of plasticity IP of Atterberg on one hand and the values obtained with the methylene blue VB on the other hand we end to the following linear correlation. This shows that the clay activity of the tested soils increases with its plasticity. This confirms the tendency of the literature but with a specific correlation with the formations in Taza urban area (Figure 5).

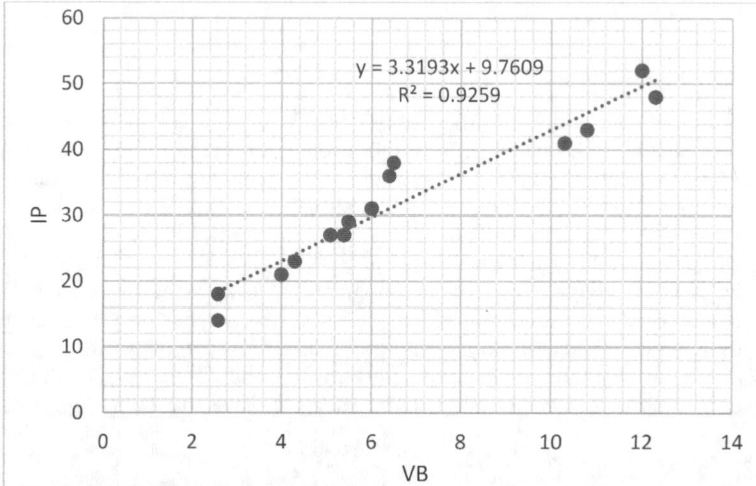

Figure 5: IP and VB Correlation of the marl- clay soils of Taza

6.3. Comparison between plasticity (IP-WL) and swelling potential

The carry forward of the plasticity index values IP according to the liquidity limits on the Casagrande diagram (Figure 6) by beach of the swelling pressure of our primarily marl and clay samples within the meaning of geotechnical classification, enable us to note that:

The evolution of the results take place in a way quasi parallel with the right-hand side of line limitation, such as $IP = 0.73\ (WL - 20)$ between the frankly argillaceous grounds and those muddy.

The cloud of the whole of couples (WL, IP) representative of the grounds fairly to strongly inflating are located in the field of index of plasticity IP higher than 20% and liquid limit WL ranging between 40% and 60%.

The swelling and shrinkage of the marl results in a loss of cohesion and consequently a high susceptibility to erosion according to Goutte-Lima et al.

6.4. Exploitation of the experiments

The marl grounds which constitute major urbanized surface or candidate to the extension of the occupation of the regional planning are often very plastic, which is in strong correlation with the Blue values reflecting their activity as an argillaceous ground classified according to the diagram (Fig. 5). Then the majority of our samples relatively dispersed on the town of Taza for a good representativeness are of swelling fairly to strongly powerful.

This predisposition and always in reference to our experimental tests and those of the literature results in considering that the marl object of our investigations are easily candidates with erosion under the action of the water streaming as soon as their cohesion has suddenly lost its value or that this one is weak right now.

The climatic conditions which lead to the marl saturation and particularly after their prolonged drying out or loss of its dried density by withdrawal, and the time as of bad weather offers this context of its cohesion loss.

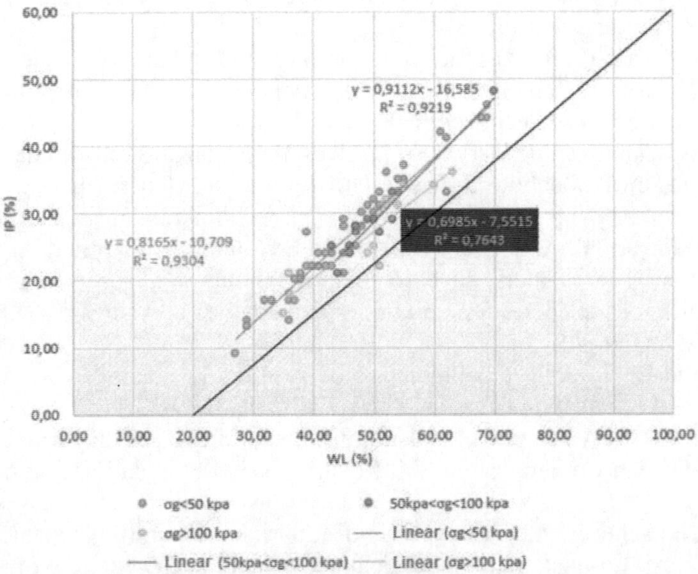

Figure 6: Correlations IP-WL according to the power of the swelling pressure of marly soils

7. Conclusions

Through the facet of our experiments and investigations by taking into account the got results we could draw up an erosion prediction methodology. Then soil loss is reported to cohesion loss with a correlation rate R^2 of 0.91 in one hand, or reported to cohesion value with a correlation rate R^2 of 0.95 in the other hand. This leads to a time saving of at least half. Which makes it possible to anticipate the risks related to the exhibitions of the marl grounds such as those of Taza with the phenomena of streaming of water. Which is possible by having recourse to the mechanical tests instead of oedometric test which needs longer time. That is all the more alarming as the marls are potentially inflating and that they are of low cohesion or that they are threatened of decoherence.

This kind of grounds, suitable for volumetric variations of expansion-withdrawal following the fluctuation of their water behavior, is reported on the cards of zoning geotechnics of the town of Taza. It will help the town planners in the construction to act in an active way against the damage by erosion of surface or mass in the form of slip.

Among the damage of erosion, we recall the obstruction of the networks of cleansing, the undermining and washing away of the foundations of buildings, works, roads and others as well as the raising of levels of high waters of believed in the wadis.

The results of the geotechnical tests of identification, swelling and mechanical parameters on the one hand, as well as those of the erosion on the other hand, reinforce our proposed method for a rigorous indirect and practical approach. It may be easily applied to other cities in the Fez-Meknes region like Taounate and My Yaacoub where marl constitutes the main bedrock.

8. References

[1] Ejjaaouani, H. (2009) Interactions des fondations et des sols gonflants : pathologie, calculs et études expérimentales. Thèse de doctorat spécialité géotechnique. Ecole nationale des ponts et chaussées de Paris. 243p.

[2] Suter, G. (1980) Carte géologique de la chaîne rifaine au1/500 000.Notes et Mémoires du Service géologique du Maroc, 245a.

[3] Leblanc, D. (1975) Etude géologique dun Rif externe oriental de Taza (Maroc).Notes et Mémoires du Service géologique du Maroc, N°281. 160p.

[4] El Qandil, M. (2007) "Recherche et valorisation des matériaux de construction et des argiles industrielles dans les provinces nord du Maroc ; possibilités d'exploitation et problèmes environnementaux. Thèse d'Etat Es Sciences Fac. Sc. Rabat. 340 p.

[5] Robillard, D. (1978) Etude pétrographique du complexe basaltique du Moyen Atlas

IOP Conf. Series: Materials Science and Engineering **353** (2018) 012010 doi:10.1088/1757-899X/353/1/012010

septentrional (région de Taza, Maroc) in thèse de 3e cycle spécialité géologie. Univ. des sciences et techniques de Lille, p173

[6] Urban Agency of Taza (2009) Diagnostic report, analysis and prospective evaluations. 317p.

[7] Mariotti, M., Naciri, H. Demehati, A. (1987) Erodabilité des sols un problème superficiel de fond.Revue marocaine de génie civil N 16. pp 12-27.

[8] Goutte-Lima M. Camapum de Carvalho, J. Alves Da Frota C (2008) Etude des caractéristiques géotechniques et d'érodablité des sols dans la ville de Manaus, Brésil. Revue française de géotechnique n°123. pp 3-13

[9] Merzouk A., Dahman H., Abdellaoui B., Houssa R. (1998) Transformation des terres du Rif et son impact sur l'envasement du barrage Ibn Battouta (Tangérois, Maroc). Congrès « La réalité de terrain en télédétection : pratiques et méthodes ». AUPELF-UREF No7, Sainte Foy PQ, Canada, pp339-348.

[10] Komornik J. and David A (1969) Prediction of swelling pressure of clays. Journal of the Soil Mechanics and Foundation Division, vol. 95, n° 1, pp. 209-225

[11] Brackley I.J.A. (1983) An empirical equation for the prediction of clay heave Proceedings of the 7th regional Asian Conference on Soil Mechanics and Foundation Engineering, vol 1. pp.8-14

[12] Chrétien, M. Marache, A. and Fabre, R. (2008). Shrinkage-swelling behavior and variability of clayey soils of Plio-quaternary geological formations in the suburbs of Bordeaux (France). International Symposium SEC 2008. Drought and construction, LCPC, France: 63-73.

[13] Chen F.H. (1973) The Basic Physical Property of Expansive Soils. Developments in Geotechnical Engineering, vol. 12, Elsevier Amsterdam Haifa, 1,pp.17–25.

[14] Kassif G and Baker R. (1971). Experimental relation between swell pressure and suction. Géotechnique, vol. 21 , N°3, pp. 245-55 Géotechnique, vol. 21, n°3, pp. 245-55

IOP Conf. Series: Materials Science and Engineering **353** (2018) 012011 doi:10.1088/1757-899X/353/1/012011

Geotechnical valorisation of large recycling concrete in Fez-Meknes region (Morocco)

A. DEMEHATI[1*], A. ABIDI[1], M. EL QANDIL[1]

[1] University Sidi Mohamed Ben Abdellah, Faculty of Sciences Dhar El Mahraz, Department of Geology, P.O. Box 1796, Fez, Morocco

E-mail*: abdelghani.demehati@usmba.ac.ma

Abstract. The growing and accelerated development of agglomerations is resulting in increasing pressure on deposits of materials as natural resources. This results in shortages of aggregates including sand. Situation in the different regions of Morocco, including that of Fez-Meknes. In addition, the threat to the stability or durability of buildings in general and road works and their annexes in particular is often started either from the surface of the grounds or from areas vulnerable to their geotechnical contact as seating materials surrounding them. The large concrete with recycles aggregates filling or protective mask provides adequate solutions. According to the results of our research, it offers a well-adapted physical and mechanical characterization in transition between conventional concretes, whether or not they are armed, and their support. Its use of protection against the effects of erosion or scouring and against the seismic movements further strengthens its potential field employment.

1. Introduction

Today, more than half of the world's 7 billion people live in urban areas, particularly cities. Current trends indicate that the same will be true by 2050 for two-thirds of the world's population. This sustained urbanization is marked by the insufficiency of the natural resources of which those called mineral in answer to the needs of constructions for the urban development [1]. Ecologists by a parallelism between urban systems and ecological systems conclude that an urban metabolism shows a strong dependence on flows of building materials [2,3,4]. This translates into the strong pressure it has on the environment to absorb construction waste or construction demolition according to their age. To better fit into a circular economy, it is necessary to develop more sustainable modes and cycles of consumption of resources. We are therefore invited to use an approach of approach by multiple uses. From this angle we seek to promote the use of recycled aggregates material through better scientific and technical knowledge. The latter obtained from the standard conventional concrete construction will be incorporated to obtain large enough resistant concrete. Among the multiple uses that we propose:

- Firstly, the confinement of the grounds [5] or embankments envelops of the foundations of construction, in particular against the seismic risk as shown by Figure 1.
- Secondly, the transition of catching of the level of sitting on support ground of better bearing, exceeding those of surfaces sensitive to swelling-withdrawal or collapsible (Figure 2).
- Thirdly, intermediate seating of structures, obstruction of karst or anthropic cavities [7] and reinforcement of various urban water networks, to mitigate or stop leaching phenomena by erosion.

IOP Conf. Series: Materials Science and Engineering **353** (2018) 012011 doi:10.1088/1757-899X/353/1/012011

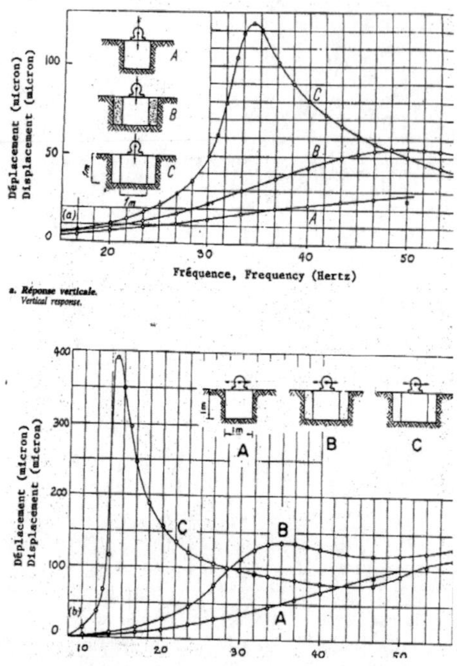

Figure 1. Vertical (a) and horizontal (b) displacement of foundation in response to seismic load (according to NOVAK) [6]

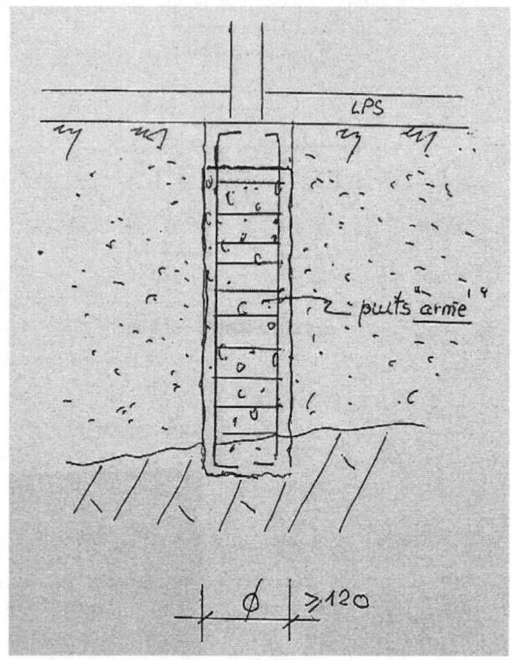

Figure 2. Semi deep large concrete foundation for light load descent

2. Research Methodology

2.1. Presentation of the approach and context

Related to Taza region, Morocco, our investigations were carried out according to the following steps:

- The contextual analysis of the use of mineral resources, sands and gravels, as well as the concretes that they compose. This directs us towards the source aggregates of calcareous origin or wadi for the sand.
- The destination analysis of a concrete based on its interaction with the soil or embankment envelops, in particular, the transition interfaces. This latter makes use of the strength or compactness of the large concrete product.
- The development of a practical approach for a better optimization of use by recycling materials from demolition concretes or waste from construction sites.

2.2. Type of large concrete envisaged

2.2.1. Natural aggregates sources

The aggregates used in the parent-concrete formulation are derived from the middle Atlas limestone quarries, more precisely Sefrou quarries [8,9] and the wadi sand as an alternative.

2.2.2. Definition of Big Concrete

Large concrete is a concrete usually dosed between 150 and 250 kg of cement with an increase in the amount and maximum diameter of pebbles and decrease in the amount of sand. The large targeted concrete is soft and homogeneous, which allows it to substantially fill the cavities at the walls of the foundation well digs manually or shovel.

3. Material and method of experimentation

In order to make the large concrete object of our research we have been careful to control its genesis as highlighted by the phases of our experiments. The natural aggregates of the city of Sefrou are the base of the concrete of the laboratory cylinders. This parent concrete is in turn considered as "artificial rock" for obtaining recycled bass. In parallel this rigor is accompanied by a simulation of concrete crushing by a heavy compactor site. Two different formulations are made. In the first sand comes from recycling. Whereas in the second it comes from wadi.

3.1. Storage of parent concrete cylinders

More than one hundred cylinders of source concrete have been selected and stored (Figure 3). They carry the same components of formulations "aggregates, sand, cement, water." according to the standard of concrete NF EN 206 [9]. Their quality is identified by tests of density, compressive strength and traction. These characteristics are: a density of 2,4-a compressive strength at 7 days of age between 16 and 21 MPa for 21 and 27 Mpa at 28 days. While the one with traction at 28 days is between 1.8 and 2.7 MPa.

Figure 3. Cylinder of parent concrete after compression failure

Figure 4. Raw recycled concrete gravels

Figure 5. Selected recycled gravel

3.2. Crushing.

The particle size desired by the crushing spread from 0.1 to 40 mm in diameter using the steam roller (Figures 4 and 5).

3.2.1.Wheel crushing

The advantage of this grinding is the obtaining of a grain size of crude enough coarse.

3.2.2. Cylinder crushing

The product of the under-roller passes of the compactor made it possible to recover a crushed crude of granulometry rich enough in fine.

3.3. Identification tests

Concrete aggregates from cylinder crushing are sorted, crushed and sieved. The physical characteristics determined in the laboratory are presented in the tables below.

Table 1. Physical characteristics of recycled aggregates (chippings and pebbles)

Characteristics of chippings and pebbles Averages	Medium
Natural water content w (%)	2.36
Actual density (g/cm3)	2.32
Real density imbibed (g/cm3)	2.48
Bulk density (g/cm3)	1.33
Atterberg limits wl, wp et Ip(%)	Atterberg limits

Table 2. Physical Characteristics of Recycled and Wadi Sand

IOP Conf. Series: Materials Science and Engineering **353** (2018) 012011 doi:10.1088/1757-899X/353/1/012011

Characteristics of sands	Medium	
	Recycled sands	Wadi sands
Natural moisture content w (%)	9.12	4.1
Actual density ρr (g/cm3)	2.01	2.62
Real density imbibed (g/cm3)	2.26	2.66
Apparent density (g/cm3)	--	1.5
Atterberg limits wl, wp et Ip (%)	Not measurable	Not measurable

The curve of the granular mixture is as follows (Figure 6).

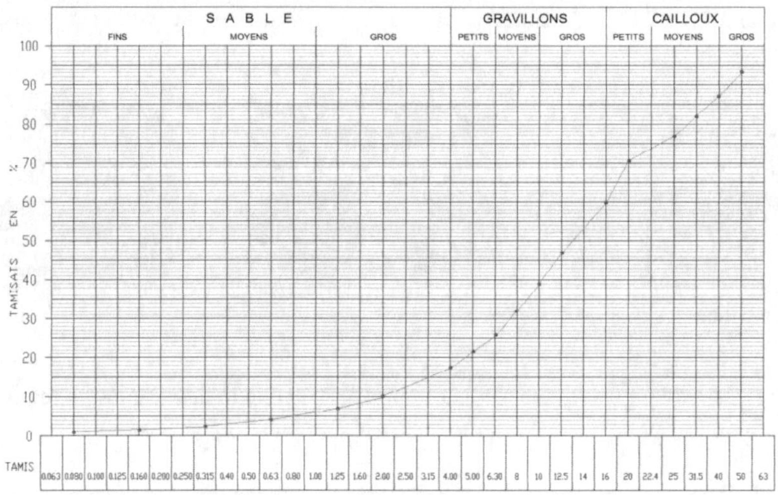

Figure 6. Particle size analysis of the mixture of recycled gravel

The Los Angeles "LA" fragmentation resistance of recycled aggregates is 29% according to the standard NM 10.1.138 [10] reference of the European standard NF EN 12620 [11]. It defines a maximum value "LA" of 30 for a good concrete of current construction.

3.4. Formulations of the large Concrete

The formulation of the large concrete is composed according to the abacuses of the usual concretes [12,13].

3.4.1. Dosage and quality of the cement

The simplified method has been used to determine the quantity of cement required in kg per m3 of fresh concrete used. The dosages are guided by the approximate resistance targeted for the concrete.
The cement is of Portland type compound CEM II 32.5 with an absolute minimum resistance of 30 MPa at 28 days according to the European standard NF EN 197-1 [14]. It is well adapted to our type of big concrete.

3.4.2. Determination of aggregates and water

The granular composition is approximated in volume for the manufacture of the experimental cylinders [15]. The entries of the following charts take into account the initial humidity of the recycled aggregates and the handling of the large fresh concrete (Figure 7).

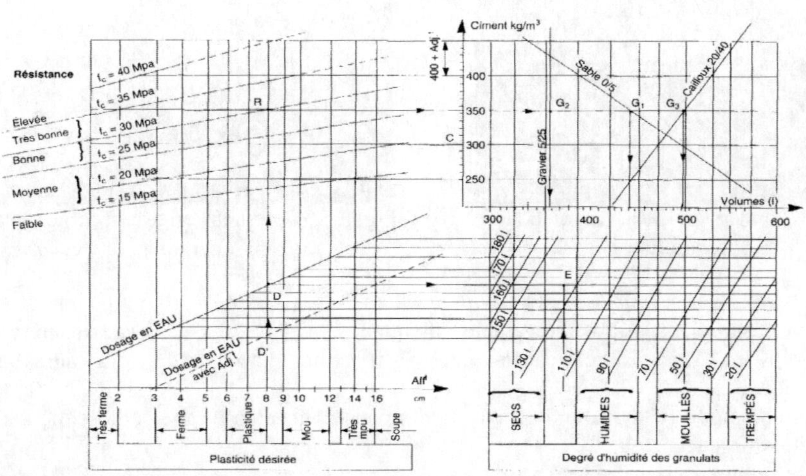

Figure 7. About assays of concrete components according to simplified method[12]

3.5. Practical application of the proposed simplified method

3.5.1. Formulation of 100% recycled aggregates

For our formulation of large concrete only recycled aggregates are retained initially: sands and gravel from our crushing to which we have added the quantities of cement and water predetermined. During this composition as for the following we have targeted a cone plasticity sag of 10 cm for a better manoeuvrability on the one hand, and a resistance of 5 Mpa to 28 days on the other hand.

3.5.2. Formulation of the big concrete with recycled gravel and wadi sand

For the second formulation we substituted the sand of wadi sand recycling by sieving the bass to 6mm. The sieve serves us for sand corrections useful to the first formulation of all recycled.

4. Results and discussion

4.1. Basic data

The large concrete made from recycled aggregates is intended for filling to improve the hydro-geotechnical context. A maximum dimension of the aggregates of 40-50 mm. A manoeuvrability of 10 cm is very convenient. The dosage cement is guided by a resistance of fc28d = 5MPa.

4.2. Simple compressive strength of concrete with 100% recycled aggregates (Figure 8)

The test consists in imposing, between two plates of a press, a constant speed of longitudinal deformation of each specimen in large concrete of 16 cm of diameter and 32 cm of height. (Figure9). The results obtained are shown in Table 3.

IOP Conf. Series: Materials Science and Engineering **353** (2018) 012011 doi:10.1088/1757-899X/353/1/012011

Figure 8. Cylinders of 100% recycled large concrete

Figure 9. Compression test of large concrete cylinders

4.3. *Dynamic auscultation tests by propagation of sound velocity (Figure 10)*

The standard measurement [16] of the velocity of sound is read across each large concrete specimen before it is crushed. This test reflects the internal structure thanks to the longitudinal waves of compression (P waves).

The device is composed of the sound propagation device connected to an oscilloscope acting as an electro-acoustic transducer

The results are summarized in Table 3 below.

Table 3: Resistance to compression results and corresponding sonic velocity at 28 days

Large concrete with 100% recycledaggregates			Large concrete with added wadi sand		
RC7d (Mpa)	RC28d	Speed (Km/s)	RC7d (Mpa)	RC28d	Speed (Km/s)
9,5	9,5	4,08	**8,5**	**10,5**	3,57
16	**12,5**	4,17	8,5	12,0	3,78
11	13,0	**4,13**	9,0	10,0	**3,75**

We note that the compressive strength of recycled concrete is nearly half that of parent concrete. In parallel its density is of the order of 2.2 lower than 2.4 parent concrete. This confirms greater porosity for the first in comparison.

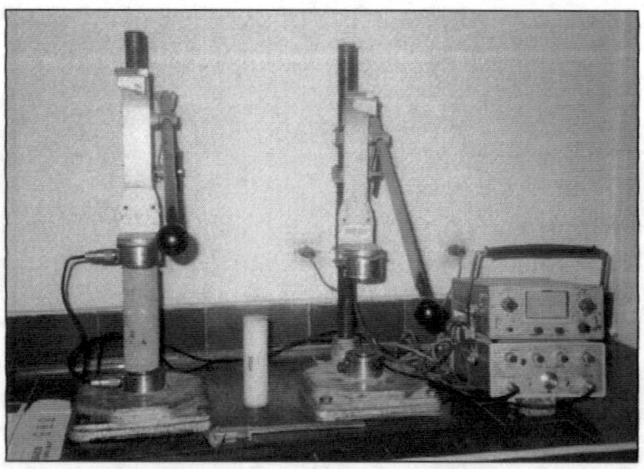

Figure 10. Ultrasonic apparatus

We also find that the values of the resistance of the large concrete are higher with the addition of the recycled sand than with the wadi sand. The median value of the resistance to 7 days of age for the first one is superior to the median value of the second at 28 days. The same trend is verified by sonic velocities performed at 28 days. We can retain a minimum resistance of 10 MPa.

IOP Conf. Series: Materials Science and Engineering **353** (2018) 012011 doi:10.1088/1757-899X/353/1/012011

Moreover, the homogeneity of the large concrete, as shown by its granulometric curve of the dry mix (Figure 6), reinforces the protection of soils and embankments of the foundation or containment against the erosive effect due to possible runoff in the absence of the large concrete.

5. Conclusions

The experiments carried out on the large concrete based on aggregates from recycled concrete can lead to a sufficiently rigid and waterproof material. It meets the needs of continuity in gradual transition interface to the support of the structures.

This type of recycling will largely contribute to the recovery of large concretes in general and in particular in the region of Fez-Meknes which is experiencing exploitation pressures of the aggregates leaving the risks of their scarcity to hover. The discharge of the concrete discharges will be reduced.

Areas of research can be developed from the need for reinforcement, protection or consolidation necessary for the stability and durability of the works with respect to geotechnical hazards or vulnerabilities. For our part, we have indicated certain uses favourable to the improvement of the soils of erosion protection and limitation of the amplitudes of seismic actions.

6. References

[1] World Urban Observatory (2007)- UN News Center. United Nations Population Fund (UNFPA) and World Bank, New York.

[2] Botkin, D.B. and Beveridge, C.E. Cities as environments. Urban Ecosystems(1997).

[3] Moffatt, S. & Kohler, N. (2008) Conceptualizing the built environment as a socialecological system. Building Research and Information, pp 248-268.

[4] Newman, P. W. G. (1999) Sustainability and cities: Extending the metabolism model.Landscape and Urban Planning, 44,pp 219-226

[5] M. Novak.(1974) Dynamic stiffness and damping of piles. Canadian Geotechnical.Journal, 11, 574-598, New York,pp 47-74.

[6] Novak, M. (1987). "State of the art in analysis and design of machine foundations."ECM. N.Y., pp 171-192.

[7] Abidi, A.& Demehati,A.& Banouni,H.& El Qandil,M.(2017) The Importance of Underground Cavities.Case of the Agglomeration of Fez (Morocco).GGE, pp 1–14

[8] Carte géologique du Maroc – Sefrou au 1/100000(1989) Editions du Service Géologique du Maroc. Notes et Mémoires N°354.

[9] El Fellah Idrissi,B.& Hinaje,S.(2017).Geo-Eco-Trop.,Characterization of the mineralization of groundwater of Sefrou and Anoceur causses (Morocco), pp 119-134

[10] Standard NM 10.1.138(1995). Aggregates - Test "Los Angeles".

[11] Standard NF EN 12620 (2003). Aggregates for concrete

[12] Dreux G. (1970), Guide pratique du béton. Collection de l'ITBTP- p257

[13] Dreux G. et Festa J. (1998).Nouveau guide du béton. Eyrolles, Paris, p409

[14] European Standard NF EN 197-1(2012).Common Cements (Official Journal of the EU)

[15] Larrard, F. (2000). Structures granulaires et formulation des bétons. - LCPC,Paris, 414 p.

[16] Standard NF EN 12504-4(2005). European standard for dynamic auscultation for concrete.

IOP Conf. Series: Materials Science and Engineering **353** (2018) 012012 doi:10.1088/1757-899X/353/1/012012

A comparative study of kalman filtering based observer and sliding mode observer for state of charge estimation

Hicham BEN SASSI[1], Fatima ERRAHIMI[1], Najia ES-SBAI[1], Chakib ALAOUI[2]

[1] Laboratory of Renewable Energies and Intelligent Systems (LERSI), Faculty of Sciences and Technology, Sidi Mohamed Ben Abdellah University, Fez,Box 2202, Fez, Morocco.
[2] EUROMED University, Fez, Morocco

hicham.1bensassi@gmail.com

Abstract. Nowadays, electric mobility is starting to define society and is becoming more and more irreplaceable and essential to daily activities. Safe and durable battery is of a great significance for this type of mobility, hence the increasing interest of research activity oriented to battery studies, in order to assure safe operating mode and to control the battery in case of any abnormal functioning conditions that could damage the battery if not properly managed. Lithium-ion technology is considered the most suitable existing technology for electrical storage, because of their interesting features such as their relatively long cycle life, lighter weight, their high energy density, However, there is a lot of work that is still needed to be done in order to assure safe operating lithium-ion batteries, starting with their internal status monitoring, cell balancing within a battery pack, and thermal management. Tasks that are accomplished by the battery management system (BMS) which uses the state of charge (SOC) as an indicator of the internal charge level of the battery, in order to avoid unpredicted system interruption. Since the state of charge is an inner state of a the battery which cannot be directly measured, a powerful estimation technique is inevitable, in this paper we investigate the performances of tow estimation strategies; kalman filtering based observers and sliding mode observers, both strategies are compared in terms of accuracy, design requirement, and overall performances.

1. Introduction

State of charge is considered one of the crucial information for Battery management system especially in electric vehicles, which have the battery as the only source of energy, however this crucial information is not easily obtained sins it is an inner stat which cannot be directly measured, hence the necessity of developing powerful estimation strategies and approaches, to accurately estimate the state of charge.

In literature the proposed estimation techniques could be divided into three main families. The first category contains the colomb-counting method [1], which is one of the earliest strategies used for state of charge estimation, it's based on the computation of the amp-hours by using load current integration, however the drawbacks of this approach, such as the unknown initial stat of charge, and its sensitivity to measurement noise made it urgent necessary to search for other strategies, such as the case of the second category which is based on closed loop state observers like kalman filtering [2], sliding mode observer[3], the third category is based on the block box models and the artificial intelligence, such as the artificial neural network [4] and fuzzy logic [5], these strategies are known for the independency of their performances on the battery model, and their capability to be used for almost all battery technologies, however in this paper we will focus on the second category strategies, particularly kalman filters and sliding mode observers, for their high performances, and their capability of performing well under unpredicted conditions.

The rest of this paper is organized as follows, in section 2 an equivalent circuit model is presented to model the battery. In section 3 the design processes of both unscented kalman filter and sliding mode observer are explained. Section 4 elaborates the simulation results. This manuscript ends with a conclusion.

2. battery modelling

Both Unscented kalman filter and sliding mode observer are model based observers and requires a prior knowledge of the battery model, hence due to its robustness in modeling the phenomenon inside the battery and its implementation simplicity, the first order thevnin model given in figure 1, is chosen in this paper. The model is composed of a shunt resistor R0, an RC branch to model polarization effect, and a nonlinear voltage source Uoc, which models the nonlinear dependency between the state of charge and the open circuit voltage. In order for the model to mimic the performance of the real battery, the internal parameters (R0, R, C), were identified using Levenberg Marquardt nonlinear least square algorithm, the parameters are given in Table (1).

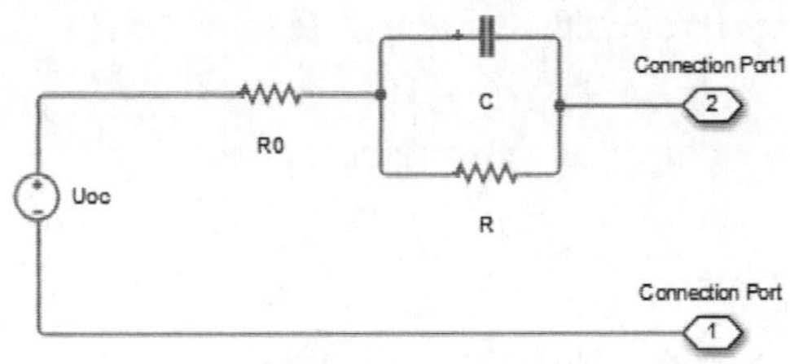

Figure 1: Lithium-ion battery model

Table 1. Battery internal parameters.

Parameter	**R_0(Ω)**	**C (F)**	**R(Ω)**
Value	0.0016	24000	0.0024

3. State of charge estimation

3.1. Kalman filters

Kalman filter (KF) are online and sometimes considered as the optimal mean to predict and correct time-varying system in a way that minimizes the mean of the squared error. Kalman filter is only valid for linear systems, for nonlinear systems, other variant of KF have been proposed through a linearization procedure, starting from extended kalman filter EKF [2] to sigma point kalman filters SPKF [6, 7]. In this paper the performances of the unscented kalman filter will be presented to illustrate the performances of kalman filters in general. The UKF is based on the unscented transformation as linearization procedure. It is a method for calculating the statistic of a random variable that undergoes a nonlinear transformation, to overcome the nonlinear variation of the OCV with respect to the state of charge. As mentioned above, KF require a dynamic state model of the battery, in order to develop a simulation model for the emulation of battery behavior, in this paper the stat space representation used for the UKF is presented by equations (1), (2). UKF algorithm is presented below, for background theory refer to [6].

$$X_{(k+1)} = f(x_k, i_k) = \begin{bmatrix} 1 & 0 \\ 0 & e^{-\frac{T}{RC}} \end{bmatrix} X_k + \begin{bmatrix} \frac{-T}{C_b} \\ \mathrm{R}(1 - e^{-\frac{T}{\mathrm{RC}}}) \end{bmatrix} I_b + wk \tag{1}$$

$$y(k) = h(x_k, i_k) = U_{oc}(SOC) - X_2(k) - R_0 i_b(k) + vk \tag{2}$$

Step 1. Initialization:

$$Q(k) = E\{w_k w^T{}_k\};\ R(k) = E\{v_k v^T{}_k\};\ \ E\{w_k v^T{}_k\} = 0 \quad (3)$$

$$P_0 = E[(x_0 - \bar{x}_0)(x_0 - \bar{x}_0)^T] \quad (4)$$

$$\bar{X}_0 = E[x_0] \quad (5)$$

Where Q(k) and R(k) are diagonal matrices that represent respectively the process noise (wk) covariance and the observation noise (vk) covariance, P_0 is the stat covariance matrix, and $\bar{X}_0$ represents the main.

Step 2. Compute sigma points

A set of weighted sigma points are deterministically computed, as shown in the following equation:

$$\mathcal{X}_{k-1} = [\bar{x}_{k-1},\ \bar{x}_{k-1} + \left(\sqrt{(n+\lambda)\boldsymbol{P}^-_{x,k-1}}\right), \bar{x}_{k|k-1} - \left(\sqrt{(n+\lambda)\boldsymbol{P}^-_{x,k-1}}\right)] \quad (6)$$

Where λ is scaling parameter that represents how far the sigma points are separated from the mean, and n is the system dimension, in our case n=2, for more background theory refer to [8].

Step 3. Prediction phase.

1) Propagating sigma points through the system/state function f(x, i_{bk}):

$$\mathcal{X}^{(i)}_{k|k-1} = f(\mathcal{X}^{(i)}_{k-1}, i_{bk}) \quad (7)$$

2) Calculating the propagated mean and covariance:

$$\bar{x}_{k|k-1} = \textstyle\sum_{i=0}^{2n} Wm.\, \mathcal{X}^{(i)}_{k|k-1} \quad (8)$$

$$\boldsymbol{P}^-_{x,k|k-1} = \textstyle\sum_{i=0}^{2n} Wc[\mathcal{X}^{(i)}_{k|k-1} - \bar{x}_{k|k-1}][\mathcal{X}^{(i)}_{k|k-1} - \bar{x}_{k|k-1}]^T + Q_{k-1} \quad (9)$$

Where Wm and Wc represents the weights of sigma points.

3) Computing the new sigma points matrix using $\bar{x}_{k|k-1}$ and $P^-_{x,k|k-1}$.

$$\mathfrak{x}_{k|k-1} = \left[\bar{x}_{k|k-1},\ \bar{x}_{k|k-1} + \left(\sqrt{(n+\lambda)\boldsymbol{P}^-_{x,k|k-1}}\right), \bar{x}_{k|k-1} - \left(\sqrt{(n+\lambda)\boldsymbol{P}^-_{x,k|k-1}}\right)\right] \quad (10)$$

4) Propagating the computed sigma points through the nonlinear function H(x, i_{bk})

$$y^{(i)}_{k|k-1} = H(\mathfrak{x}^{(i)}_{k|k-1}, i_{bk}) \quad (11)$$

5) Calculating the mean of the output variable.

$$\bar{y}_{k|k-1} = \textstyle\sum_{i=0}^{2n} Wm.\, \mathbf{y}^{(i)}_{k|k-1}; \qquad \text{For } i=1...2n \quad (12)$$

Step 4.Measurement update

1) Calculating the covariance of the measurement vector and the cross covariance respectively.

$$P_{y.k} = \textstyle\sum_{i=0}^{2n} Wc[y^{(i)}_{k|k-1} - \bar{y}_{k|k-1}][y^{(i)}_{k|k-1} - \bar{y}_{k|k-1}]^T \quad (13)$$

$$P_{xy.k} = \textstyle\sum_{i=0}^{2n} Wc[\mathfrak{x}^{(i)}_{k|k-1} - \bar{x}_{k|k-1}][y^{(i)}_{k|k-1} - \bar{y}_{k|k-1}]^T \quad (14)$$

2) Calculating the Kalman gain, in order to correct the estimations

$$K_k = P_{xy.k}.\, \mathbf{P}^{-1}_{y.k.} \quad (15)$$

3) Update the estimated Stats with kalman gain.

$$\hat{X}_k = \hat{X}_{k|k-1} + K_k(y_k - \bar{y}_{k|k-1}) \quad (16)$$

4) Update the propagated state covariance using kalman gain.

$$\hat{P}_k = \boldsymbol{P}^-_{x.k|k-1.} - K.P_{y.k}.K^T) \quad (17)$$

3.2 Sliding mode observer

Sliding mode observer SMO [9] is a robust nonlinear state estimator that has the same principal as the sliding mode control, it is widely known for the capability of compensating the effects of nonlinearity and uncertainty in battery models. This observer relies on the determination of the appropriate feedback

switching gain associated with system uncertainty bound, which balances between the chattering levels, and the error convergence speed, to achieve better SOC estimation accuracy.
Based on figure1 the dynamic state model used in this section is given by equations (18, 19, 20).

$$\dot{V_t} = -b1Vt + b1U_{oc}(\text{SOC}) - a1.I_b + \Delta f1 \quad (18)$$
$$\dot{SOC} = b2Vt - b2U_{oc}(\text{SOC}) + b2V_{RC} + \Delta f2 \quad (19)$$
$$\dot{V_{RC}} = -b1V_{RC} + a2.I_b + \Delta f3 \quad (20)$$

Where $b1 = 1/(R.C)$, $b2 = 1/(R_0.Cb)$, $a1 = \frac{R_0}{R.C} + \frac{K}{C_b} + \frac{1}{C}$, $a2 = \frac{1}{C}$, and $\Delta f1$, $\Delta f2$, $\Delta f3$ are modelling uncertainties.

The above equations are obtained by considering that there exists a piecewise linear relationship between the OCV and the SOC in a certain range of the SOC. Therefore, the OCV is expressed as a function of the SOC using piecewise linearization method.

$$U_{oc}(\text{SOC}) = \text{K. SOC} + \text{v} \quad (21)$$

Where the values of K and v are constants in every 10% of SOC, which allow us to represent the time derivative of U_{oc} in each 10% of SOC.

$$Uoc'(SOC) = -K\frac{I_b}{Cb} \quad (22)$$

In order to accurately estimate the state of charge, the SMO equations are proposed as follows.

$$\dot{\hat{V_t}} = -b1\hat{V}t + b1U_{oc}(\widehat{\text{SOC}}) - a1.I_b + \hat{\Gamma}_1 sgn(e_{Vt}) \quad (23)$$
$$\dot{\widehat{SOC}} = b2\hat{V}t - b2U_{oc}(\widehat{\text{SOC}}) + b2\hat{V}_{RC} + \hat{\Gamma}_2 sgn(e_{Uoc}) \quad (24)$$
$$\dot{\hat{V}}_{RC} = -b1\hat{V}_{RC} + a2.I_b + \hat{\Gamma}_3 sgn(e_{VRC}) \quad (25)$$

Where

$$sgn(e) = \begin{cases} +1, & e > 0 \\ -1, & e < 0 \end{cases} \quad (26)$$

And stats error is

$$e_{Vt} = Vt - \hat{V}_t \quad (27)$$
$$e_{Uoc} = U_{oc}(\text{SOC}) - U_{oc}(\widehat{\text{SOC}}) = Ke_{SOC} \quad (28)$$
$$e_{V_{RC}} = V_{RC} - \hat{V}_{RC} \quad (29)$$

The dynamics of the error $\dot{e}_{Vt}$, $\dot{e}_{Uoc}$, $\dot{e}_{VRC}$ are obtained by subtracting equations. (18, 19, 20) from equations. (23, 24, 25).

$$\dot{e}_{Vt} = -b1e_{Vt} + b1e_{Uoc} + \Delta f1 - \hat{\Gamma}_1 sgn(e_{Vt}) \quad (30)$$
$$\dot{e}_{SOC} = b2e_{Vt} - b2Ke_{SOC} + b2e_{V_{RC}} + \Delta f2 - \hat{\Gamma}_2 sgn(e_{SOC}) \quad (31)$$
$$\dot{e}_{V_{RC}} = -b1e_{V_{RC}} + \Delta f3 - \hat{\Gamma}_3 sgn(e_{V_{RC}}) \quad (32)$$

Based on the Lyapunov stability theory, the asymptotic convergence of the terminal voltage error can be proved by choosing the following candidate of Lyapunov function.

$$V_{V_t} = \frac{1}{2}e_{Vt}^2 \quad (33)$$

To guarantee the existence of sliding regime, the time derivative of the candidate of Lyapunov function must be negative, and after developing the expressions of $\dot{V}_{V_t}$, the value of $\Gamma 1$ that guarantees $\dot{V}_{V_t} < 0$ is $\Gamma 1 \gg \Delta f1$. Once the sliding surface is reached then $\dot{e}_{Vt} = e_{Vt} = 0$, and the unmeasurable SOC error can be derived by replacing $\dot{e}_{Vt}$ and e_{Vt} with zeros in equation (30).

$$e_{SOC} = \{\left(\frac{\Gamma 1}{Kb_1}\right) sgn(e_{Vt})\}_{eq} \quad (34)$$

Similarly $e_{V_{RC}}$ can be derived from equation. (31) by replacing $\dot{e}_{Uoc}$ and e_{Uoc} with zero.

$$e_{V_{RC}} = \{(\frac{\Gamma 2}{b2})sgn(\{(\frac{\Gamma 1}{Kb_1})sgn(e_{Vt})\}_{eq})\}_{eq} \tag{35}$$

Finally the SMO observer is obtained by substituting equations. (34, 35) into equations. (24, 25).

4. Simulation and results discussion

Following the design processes, both sliding mode observer and unscented kalman filter were tested and simulated under matlab Simulink environment for a li-ion battery of 2Ah capacity. Simulation results are presented in figure 2, 3, 4 and 5.

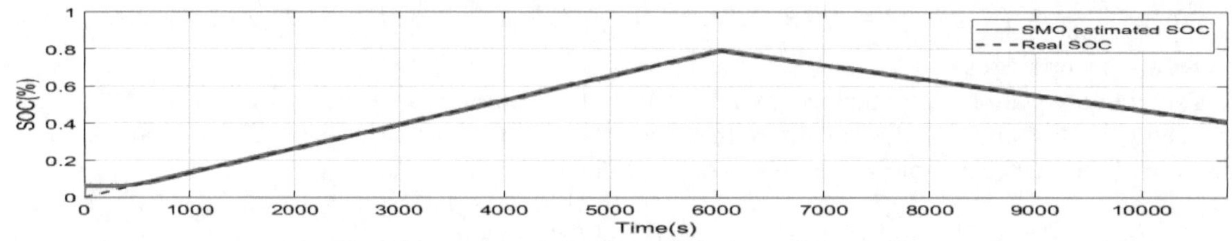

Figure 2: State of charge estimation using SMO

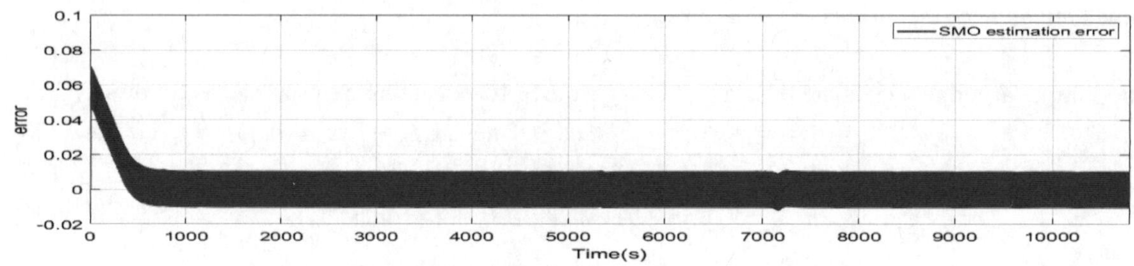

Figure 3: State of charge estimation error using SMO

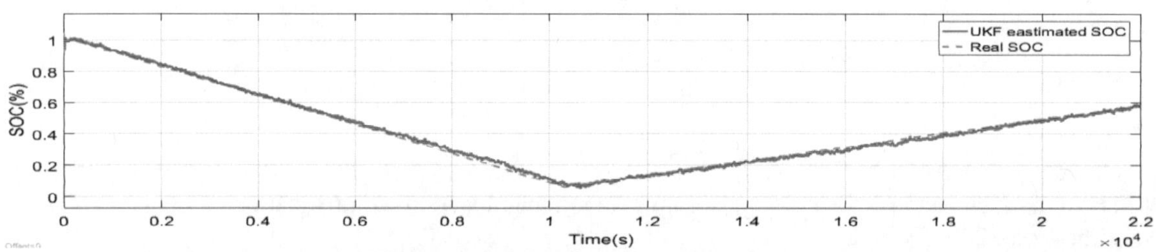

Figure 4: SOC estimation using UKF

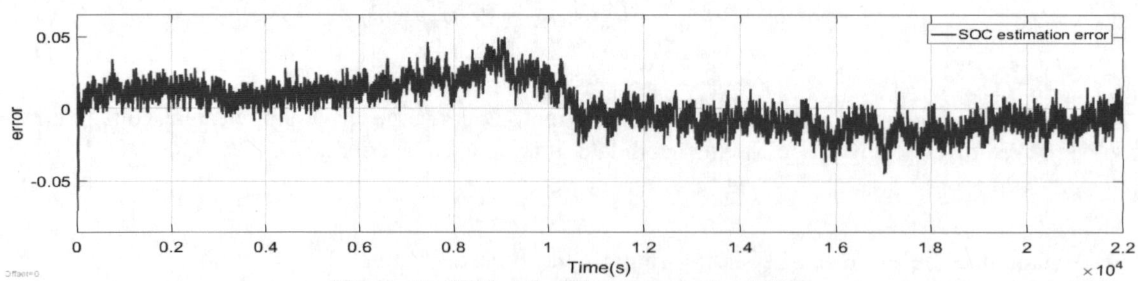

Figure 5: SOC estimation error using UKF

From the above simulation results, it is clear that both strategies perform well in SOC estimation, although the sliding mode observer (SMO) is slightly more accurate with maximal error value less than 7%, at the beginning of the estimation, and mean error less than 1.7%, compared to maximal error value less than 6% and mean error less than 2% in the case of UKF, which is due to the robustness of the SMO against modeling errors and internal parameters uncertainties. On the other hand it is clear that SMO suffers from high frequency chattering around the true SOC, which blurs the SOC estimation and affects the stability of the observer, and unlike UKF, SMO has a relatively slow convergence time.

5. Conclusion

The accurate estimation of stat of charge, is a key factor of managing batteries with high efficiency. In this paper, a comparative study of kalman filters and the sliding mode observer was carried out, from design challenges to estimation accuracy. From the obtained results, it can be concluded that both strategies are competitive in terms of precision, the SMO is more robust against model uncertainties if the feedback gain is correctly chosen. From implantation point of view and based on work presented in [10, 11], the UKF requires a powerful platform, on the other hand the SMO is easy to implement since it only requires the computation of the feedback gain.

References

[1] Kong Soon Ng , Chin-Sien Mooa, Yi-Ping Chen, Yao-Ching Hsieh, "Enhanced coulomb counting method for estimating state-of-charge and state-of-health of lithium-ion batteries" Applied Energy Volume 86, Issue 9, September 2009, Pages 1506-1511

[2] H.S. Ramadan , M. Becherif, F. Claude "Extended kalman filter for accurate state of charge estimation of lithium-based batteries: a comparative analysis " international journal of hydrogen energy 42 (2017) 29033 -29046.

[3] Xiaopeng Chen, Weixiang Shen, Mingxiang Dai, Zhenwei Cao, Jiong Jin, and Ajay Kapoor "Robust Adaptive Sliding Mode Observer Using RBF Neural Network for Lithium-ion Battery State of Charge Estimation in Electric Vehicles" 0018-9545 (c) 2015.

[4] Wei He, Nicholas Williard, Chaochao Chen, Michael Pecht "State of charge estimation for Li-ion batteries using neural network modeling and unscented Kalman filter-based error cancellation" Electrical Power and Energy Systems 62 (2014) 783–791

[5]] W. X. Shen, C. C. Chan, E. W. C. Lo, and K. T. Chau, "Adaptive neuro-fuzzy modeling of battery residual capacity for electric vehicles," IEEE Transactions on Industrial Electronics, vol. 49, pp. 677-684, 2002.

[6] Ahmed Gaga, Hicham Benssassi, Fatima Errahimi, Najia Es-Sbai " Battery State of Charge Estimation Using an Adaptive Unscented Kalman Filter for Photovoltaics Applications" International Review of Automatic Control (IREACO) Vol 10, No 4 (2017).

[7] Gregory L. Plett "Sigma-point Kalman filtering for battery management systems of LiPB-based HEV battery packs: Part 2: Simultaneous state and parameter estimation" Journal of Power Sources 161 (2006) 1369–1384.

[8] S. Julier et J. Uhlmann, «New extension of the Kalman filter to nonlinear systems,» International Symposium on Aerospace/Defense Sensing,Simulation, and Controls, pp. 182-193, 1997.

[9] GONG Xiaohui, Huang Zhiwu, LI Liran, LU Honghai, LIU Sanfeng, WU Zhihui"A New State of Charge Estimation for Lithium-ion Battery Based on Sliding-Mode Observer and Battery Status" Proceedings of the 35th Chinese Control Conference July 27-29, 2016, Chengdu, China.

[10] Noemi Otero, Habiballah Rahimi-Eichi, Juan J.Rodriguez-Andina "FPGA implementation of an observer for state of charge estimation of lithium-Polymer Batteries" 2014 International Conference on Mechatronics and Control (ICMC) July 3 - 5, 2014, Jinzhou, China 978-1-4799-2538-4/14/$31.00©2014IEEE.

[11] Anne-Marie Lienhardt, Guillaume Gateau, and Thierry A. Meynard. "Digital Sliding-Mode Observer Implementation Using FPGA" IEEE TRANSACTIONS ON INDUSTRIAL ELECTRONICS, VOL. 54, NO. 4, AUGUST 2007. 10.1109/TIE.2007.898305.

Study of marbles from Middle Atlas (Morocco): elemental, mineralogical and structural analysis

S Khrissi[1], L Bejjit[1], M Haddad[1,*], C Falguères[2], S Ait Lyazidi[1], M El Amraoui[1]

[1] Laboratoire de Spectrométrie des matériaux et archéomatériaux (LASMAR), Unité de recherche associée au CNRST (URAC11), Université Moulay Ismail, Faculté des sciences, B. P. 11201, Zitoune, 50000, Meknès, Maroc.
[2] Département Homme et Environnement, Muséum National d'Histoire Naturelle, UMR7194, 75 013 Paris, France.

*E-mail: mhaddad22@yahoo.fr

Abstract. A series of marbles sampled from the region of Middle Atlas (Morocco), are characterized by different complementary spectroscopic techniques. X-Ray fluorescence is used to determine elemental composition of rock while X-Ray diffraction and the Raman spectroscopy are used to determine major crystalline phases (calcite and dolomite) and minor ones (quartz).The samples display typical EPR spectra of Mn^{2+} in calcite and reveal the presence of Fe^{3+} ions.

Keywords: Marbles; Middle Atlas (Morocco); characterization; XRF; XRD; Raman; EPR; Provenance.

1. Introduction

Marble is a naturally occurring stone mainly used in the construction industry or as a raw material for artwork. Since antiquity, it is one of the most frequently used materials for architecture and sculpture and therefore has been given an exceptional importance, particularly under the Romans [1,2]. Marble consists of calcite ($CaCO_3$) and dolomite ($CaMg(CO_3)_2$) grains interbedded with other minerals like quartz, mica, graphite, iron oxides, etc., which represent impurities [1,2].

The use of limestone and marble in the Moroccan structures and buildings goes back to Roman antiquity [3,4]. The ornamental rocks are abundant in central Morocco, this is the principal reason why marbles industry is more developed in this country [3,4].The goal of this work is to provide a multi-technical characterization of marbles and limestones, sampled from Middle Atlas quarries (Morocco). The characterization combines elemental, mineralogical and structural analyses.

2. Materials and characterization techniques

2.1. Samples

Several samples of marbles (fig.1 and table 1) were collected from quarries located in the region of Middle Atlas: Tiskram (KB and KN), Sidi Lamine (K1, K2, K3, K4 and K6) and Sidi Amer (K5 and K7); some of these areas contain fossils. The samples are listed in Table 1 with the corresponding colors and sites.

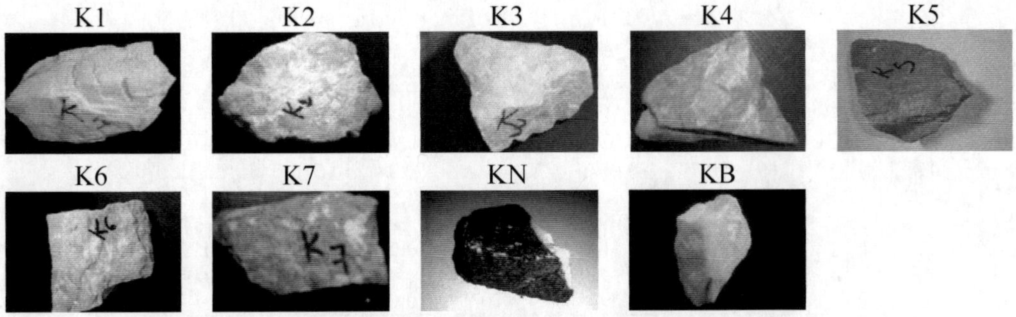

Figure 1. The samples of studied marbles.

Table 1. Samples codes and quarries of their provenance.

Sample code	Color	Provenance
K1	White	Sidi Lamine
K2	White	Sidi Lamine
K3	White/Yellow	Sidi Lamine
K4	White	Sidi Lamine
K5	Grey	Sidi Amer
K6	White	Sidi Lamine
K7	Grey	Sidi Amer
KN	Black	Tiskram
KB	White	Tiskram

2.2. Experimental techniques
Elemental analyses were determined using an Axios XRF spectrometer with dispersion wavelength. The samples were ground in an agate mortar in pellet form.

X-Ray powder diffraction analysis was carried out on powdered samples at room temperature using the Shimadzu XDR-6100 with Cu Kα radiation at the wavelength 1.5405 Å. Samples were analyzed from 5° to 70°.

Micro Raman spectra were recorded in the range of 100 - 2000 cm^{-1} using a Renishaw RM 1000 spectrometer, equipped with a CCD detector and an external Leica DMLM confocal microscope with 10x, 20x, 50x, and 100x objectives. The excitation source is a He-Ne laser operating at 632.8 nm with a resolution greater than 2 cm^{-1}. The samples were used directly, without any preparation or pre-size fraction.

X-band EPR spectra of the samples were recorded by a Bruker spectrometer at room temperature, the microwave frequency was 9.5 GHz with a 100 KHz modulation. EPR spectroscopy involves the resonant absorption of microwave power by a few milligrams of samples (≈ 200 mg) in powder form placed in a cavity inside a quartz tube, across which a magnetic field (5- 8000 Gauss) is scanned. It detects paramagnetic atoms and ions, unpaired electrons and free radicals.

3. Results and discussion

3.1. Elemental analyses: XRF
X-ray fluorescence spectrometry has been applied on powders of marbles to determine their chemical composition. The obtained results are summarized in Table 2, they show that the studied marbles contain CaO principally and small amounts of Fe_2O_3, SiO_2, MgO, Al_2O_3, Na_2O, SrO, MnO_2, Cl, SO_3, I, P_2O_5, Y_2O_3 and K_2O. All the samples contain CaO (50.1 - 55.1%) as major element and MgO (0.45 - 1.13%), SiO_2 (0.41 - 8.48%), Fe_2O_3 (0.06 – 1.35 %) and Al_2O_3 (0.2 – 1.13%) as minor elements. The analysis revealed also the presence of trace of other elements such as Na, Sr, Cl, Mn, I, S, P, K and Y in certain marbles.

Table 2. XRF analysis of marble samples.

	CaO	Fe_2O_3	SiO_2	MgO	Al_2O_3	Na_2O	SrO	MnO_2	Cl	SO_3	I	P_2O_5	K_2O	Y_2O_3	LOI*
K1	51.5	1.35	0.76	0.5	0.43	0.14	0.11	0.08	0.07	0.06	0.03	0.02	-	-	45
K2	54.8	0.13	1.54	0.62	0.34	0.12	0.03	-	-	0,03	0,03	0,02	-	-	42.3
K3	55.1	0.09	0.97	0.45	0.28	0.08	0.03	-	-	0.03	0.03	0.03	-	-	42.9
K4	53.2	0.21	2.45	0.68	0.78	0.07	0.03	-	0.04	0.03	0.02	0.15	0.08	-	42.2
K5	55.1	0.07	0.52	0.61	0.27	0.10	0.03	-	0.06	0.06	0.02	0.13	-	-	43
K6	52.6	0.28	2.48	0.85	0.99	0.09	0.03	-	-	0.04	0.02	0.06	0.12	-	42.4
K7	54.9	0.06	0.41	0.63	0.20	0.11	0.05	-	0.05	0.07	0.03	0.22	-	-	43.3
KN	50.1	0.37	5.07	1.43	1.13	0.19	0.12	0.06	-	0.3	0.02	0.08	0.12	0.004	41
KB	52	1.07	8.48	0.58	0.37	0.09	0.24	0.07	0.05	0.03	0.02	0.01	-	0.004	37

**LOI: Loss on ignition*

3.2. Crystallines phases: XRD

The various Bragg reflections are indexed using (JCPDS) files [5]. The marbles K1 and K6 are calcitic relatively pure; K2, K3, K4, K5, K7 and KB are calcitic also but with low contents of quartz. The sample KN is calcitic with low contents of both quartz and dolomite.

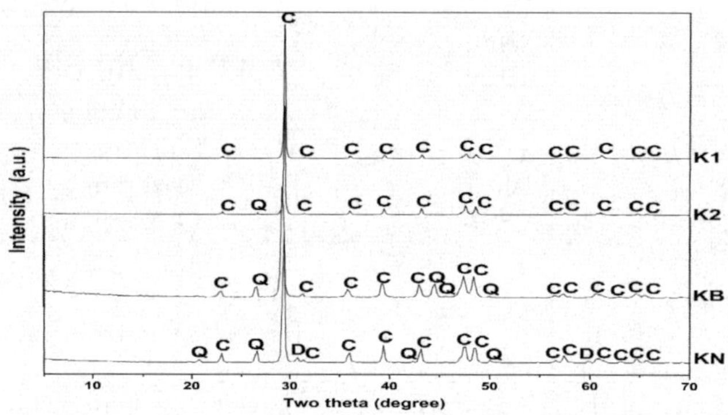

Figure 2. Powder XRD patterns of samples KN, KB, K2 and K1.
C: Calcite, D: Dolomite and Q: Quartz.

3.3. Crystallines phases: Micro Raman

The results obtained with XRD are confirmed by the Raman analyses. Indeed, the Raman spectra of marbles are similar; they show bands at 1085, 1435, 710 and 280 cm^{-1} characteristics of CO_3 groups in calcite [6-8]; the weak line at 1747 cm^{-1} may be attributed to the combination of the vibrational modes $\upsilon 1$ and $\upsilon 4$ of the CO_3.

Besides these bands, an additional wide signal at 460 cm^{-1}, signature of α-quartz SiO_2 [6,8,9], is observed in the case of the samples K5 and KN. The spectra of these two marbles also show two supplementary bands at 1325 and 1609 cm^{-1}, which are related to carbonaceous matter (graphite) in the samples [8-11]. Marbles often contain carbonaceous matter resulting in a more or less grey color [10,11].

IOP Conf. Series: Materials Science and Engineering **353** (2018) 012013 doi:10.1088/1757-899X/353/1/012013

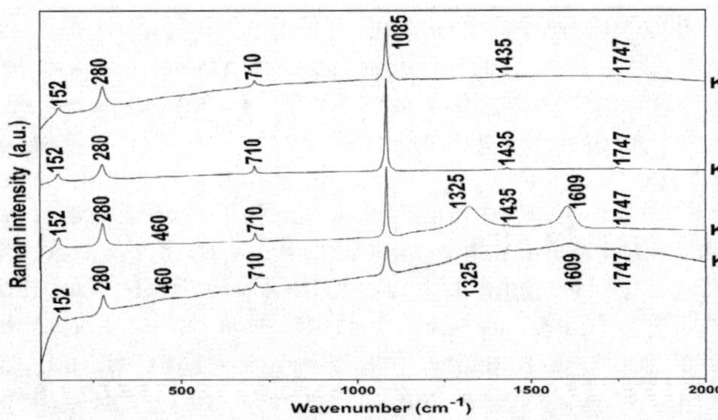

Figure 3. Raman spectra of samples K5, KN, K4 and K1.

The strongest band observed on all spectra obtained from the studied marbles corresponds to the symmetric stretching υ1 mode located around 1085 cm^{-1} and the weak lines located near 1435 and 710 cm^{-1} can be attributed respectively to the υ3 (asymmetric stretching) and υ4 (in-plane-bending) vibrational modes [12,13]. The observed frequency shifts about 152 and 280 cm^{-1} originate from the external vibrations of the CO_3 groups involving translatory and rotatory oscillations [12-14].We observe also a weak band at 1747 cm^{-1} which may be regarded as the combination band of υ1 and υ4 [12,14].

3.4. EPR spectrometry

Because of its high sensitivity, the EPR technique has been used to detect crystal imperfections that may be related to the nature of the quarry from which the marble comes; it is one of the best methods to determine the provenance of marbles. The recorded EPR spectra (fig. 4) were attributed to Mn^{2+} ions; traces of these ions are omnipresent impurities in calcium carbonates. As it possesses a $3d^5$ electronic configuration, it has six oxygen atoms as its nearest neighbours, each atom belongs to a different CO_3^{2-} group, in an octahedral structure [15-17].The Mn^{2+} presents a spectrum with six principal lines, due to the hyperfine interaction (allowed transitions), with supplementary intermediate weak lines (forbidden transitions).

The shape of the EPR features, showing dependence on the sampling quarry (fig. 4), constitutes a powerful comparison element in the provenance study of marbles. It's linked to the details of the chemical environment of the Mn^{2+} [18-20]. Indeed, in calcite ($CaCO_3$) manganese ions substitute for calcium atoms in the calcium carbonate lattice generating thus only one kind of paramagnetic site [15,18].

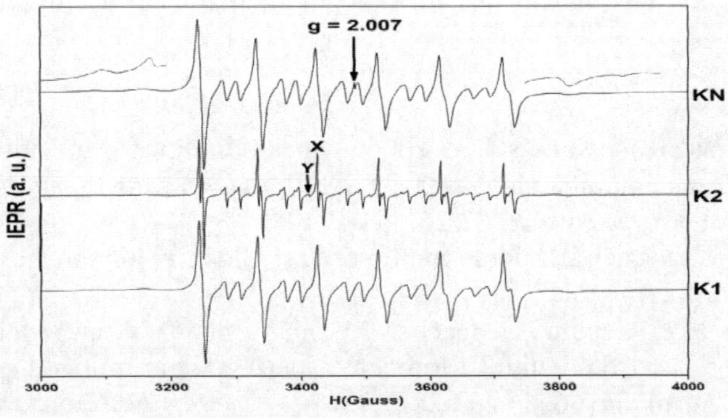

Figure 4. EPR spectra of samples K1, K2 and KN.

All EPR lines from studied marbles correspond to Mn^{2+} in Ca^{2+} sites (calcite), the other lines associated to Mn^{2+} in Mg^{2+} sites (dolomite) depend strongly on the original quarry; the dolomite/calcite ratio calculated from the EPR spectra, Y and X (fig.4) can be used as an indicator of provenance [9,21,22], it varies from one quarry to another. The other transitions are also observed in these spectra but they are very weak (fig. 4). These lines, outside the six lines of Mn^{2+}, indicate that the symmetry is slightly distorted octahedron. This is due to the presence of Mn^{2+} in a low symmetry environment caused by the random orientations in the polycrystalline powder [9,15,23].

The marble from Tiskram (KN), exhibits a weak EPR signal between the third and fourth Mn^{2+} lines at $g \approx 2.007$; (fig. 4). This signal has been attributed to a paramagnetic defect centers induced in minerals by a low level of ionizing radiation that originates from natural radioactivity. The signal intensity is an indicator of provenance and may also be useful for marble dating [9,21,24].

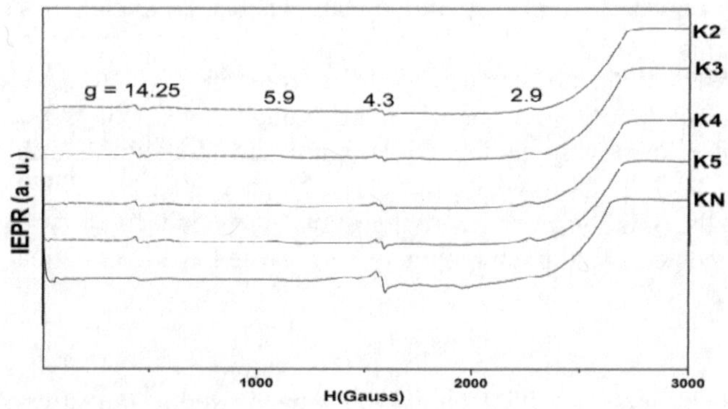

Figure 5. EPR spectra of Fe^{3+} ions in KN, K5, K4, K3 and K2 marbles.

In the low magnetic field region (fig. 5), the peaks at g≈14.25, g≈5.9 and g≈2.9 are attributed to Fe^{3+} in the calcium carbonate lattice. The peak at g≈4.3 is related to Fe^{3+} in silicates and oxides [20,21,25].

4. Conclusion

Marbles sampled at different quarries in the region of Middle Atlas (Morocco) were characterized by taking a multi-technical approach. XRD showed that the calcite is the major phase in all the marbles. These results were confirmed by Raman spectrometry which shows also the graphitic nature of marbles from certain sites. Quartz and dolomite with low content have been detected in some marbles.

The EPR study shows that all the marbles exhibit principally the same spectrum due to Mn^{2+}, but for each sample the spectrum present a shape and supplementary EPR lines which depend on the quarry origin of the marble. That means that EPR technique may be a useful method to determine the provenance of the marbles.

References

[1] Vagenas N V and Kontoyannis C G 2003 A methodology for quantitative determination of minor components in minerals based on FT-Raman Spectroscopy the case of calcite in dolomitic marble *Vib. Spectrosc.* **32** 261-64.
[2] Sarıcı D E 2016 Thermal deterioration of marbles: Gloss, color changes *Constr. Build. Mater.* **102** 416-21.
[3] Origlia F, Gliozzo E, Gandin A, Meccheri M, Spangenberg J E and Memmi I T 2012 Marbles and carbonate rocks from central Morocco: a petrographic, mineralogical and geochemical study *Environ. Earth Sci.* **66** (1) 209-22.
[4] Bouhaouli A, Caraman V and Hilali E 1979 Les marbres du Maroc *Mines, Géologie et Énergie,* **45** 35-102.
[5] JCPDS ASTM card NOS. 5-0586, 36-0426 and 33-1161.

[6] Koralay T and Kılınçarslan S 2016 A multi-analytical approach for determining the origin of the marbles in Temple-A from Laodicea ad Lycum (Denizli-Western Anatolia, Turkey) *J. Cult. Herit.* **17** 42-52.
[7] Rutt H N and Nicola J H 1974 Raman spectra of carbonates of calcite structure *J. Phys. C: Solid State Phys.* **7** (24) 4522-28.
[8] Khrissi S, Haddad M, Bejjit L, Ait Lyazidi S, El Amraoui M and Falguères C 2017 Raman and XRD characterization of Moroccan Marbles *IOP Conf. Ser.: Mater. Sci. Eng.* **186** 012028.
[9] Khrissi S, Haddad M, Bejjit L, Ait Lyazidi S, El Amraoui M and Falguères C 2017 Étude spectrométrique de marbres du Maroc central *L'Anthropologie* **121** 55-62.
[10] Sˇtastná A, Prikryl R and Jehlicka J 2009 Methodology of analytical study for provenance determination of calcitic calcite-dolomitic and impure marbles from historical quarries in the Czech Republic *J. Cult. Herit.***10** 82-93.
[11] Jehlička J, Šťastná A and Přikryl R 2009 Raman spectral characterization of dispersed carbonaceous matter in decorative crystalline limestones *Spectrochim. Acta, Part A* **73** 404-09.
[12] Gunasekaran S, Anbalagan G and Pandi S 2006 Raman and Infrared spectra of carbonates of calcite structure *J. Raman Spectrosc.* **37** 892-99.
[13] Krishnamurti D 1957 *Proc. Ind. Acad. Sci.* 27A 183.
[14] Gunasekaran S and Anbalagan G 2008 Spectroscopic study of phase transitions in natural calcite mineral *Spectrochim. Acta, Part A* **69** 1246-51.
[15] Gunasekaran S and Anbalagan G 2008 Optical absorption and EPR studies on some natural carbonate minerals *Spectrochim. Acta, Part A* **69** 383-90.
[16] Shepherd R A and Graham W R M 1984 EPR of Mn^{2+} in polycrystalline dolomite *J. Chem. Phys.* **81** (12) 6080-84.
[17] Wildeman T R 1970 The distribution of Mn^{2+} in some carbonates by electron paramagnetic resonance *Chem. Geol.* **5** (3) 167-177.
[18] Duliu O G and Velter-Stefanescu M 2006 Numerical taxonomy: Mn^{2+} EPR line shape as a criterion in provenance studies *J. Optoelectron. Adv. Mater.* **8** (4) 1557-61.
[19] Lloyd R V, Smith P W and Haskell H W 1985 Evaluation of the manganese ESR method of marble characterization *Archaeometry* **27** (1) 108-116.
[20] Polikreti K, Maniatis Y, Bassiakos Y, Kourou N and Karageorghis V 2004 Provenance of archaeological limestone with EPR spectroscopy: the case of the Cypriote-type statuettes *J. Archaeol. Sci.* **31** (7) 1015-28.
[21] Polikreti K and Maniatis Y 2002 A new methodology for the provenance of marble based on EPR spectroscopy *Archaeometry* **44** (1) 1-21.
[22] Attanasio D and Platania R 2000 ESR spectroscopy as a tool for Identifying joining fragments of antique marbles: the example of a pulpit by Donatello and Michelozzo *J. Magn. Reson.* **144** (2) 322-9.
[23] Cordischi D, Monna D and Segre A L 1983 ESR analysis of marble samples from Mediterranean quarries of archaeological interest *Archaeometry* **25** (1) 68-76.
[24] Maniatis Y and Mandi V 1992 Electron paramagnetic resonance signals and effects in marble induced by working *J. Appl. Phys.* **71** (10) 4859-67.
[25] Polikreti K and Maniatis Y 2004 Distribution changes of Mn^{2+} and Fe^{3+} on weathered marble surfaces measured by EPR spectroscopy *Atmosph. Environ.* **38** (22) 3617-24.

IOP Conf. Series: Materials Science and Engineering **353** (2018) 012014 doi:10.1088/1757-899X/353/1/012014

A technological review on electric vehicle DC charging stations using photovoltaic sources

Cheddadi Youssef[1,*], Errahimi Fatima[1], Es-sbai najia[1], Alaoui Chakib[2]

[1] Sidi Mohamed Ben Abdellah University, Laboratory of Renewables Energy and Intelligent Systems, Road Imouzzer, BP 2202, Fez, Morocco
[2] INSA, Euro-Mediterranean University of Fez, Morocco

[*] youssef.cheddadi@usmba.ac.ma

Abstract. Within the next few years, Electrified vehicles are destined to become the essential component of the transport field. Consequently, the charging infrastructure should be developed in the same time. Among this substructure, Charging stations photovoltaic-assisted are attracting a substantial interest due to increased environmental awareness, cost reduction and rise in efficiency of the PV modules. The intention of this paper is to review the technological status of Photovoltaic–Electric vehicle (PV-EV) charging stations during the last decade. The PV-EV charging station is divided into two categories, which are PV-grid and PV-standalone charging systems. From a practical point view, the distinction between the two architectures is the bidirectional inverter, which is added to link the station to the smart grid. The technological infrastructure includes the common hardware components of every station, namely: PV array, dc-dc converter provided with MPPT control, energy storage unit, bidirectional dc charger and inverter. We investigate, compare and evaluate many valuable researches that contain the design and control of PV-EV charging system. Additionally, this concise overview reports the studies that include charging standards, the power converters topologies that focus on the adoption of Vehicle-to grid technology and the control for both PV–grid and PV standalone DC charging systems.

Keywords. PV-grid charging station, Electric vehicle charging, smart grid, Vehicle to grid, Bidirectional DC converter, Energy storage unit.

1. Introduction

For economic development of any nation, the population depends strongly on fossil fuels, particularly for transportation and electricity generation. Since the number of Vehicles increases every day, the air quality become a serious problem in urban area due to the big amount of the burning of fossil fuels that are greatly responsible for global warming.

In the last few years there has been considerable interest in Electric Vehicles (EVs) and Plug-in Hybrid Electric Vehicles (PHEV), which could play an important role in reducing greenhouse gas (GHG) emissions from the transport sector, and have potential as a future alternative to internal combustion (IC) vehicles. However, meaningful GHG emissions reductions with EVs are conditional on low-carbon on their source of electric energy. According to reference [1], the life cycle GHG emissions from PHEVs is assessed and find that they reduce GHG emissions by 32% compared to conventional vehicles.

In this paper, the term of "electric vehicle" refers to all kind of transportation that contains rechargeable batteries; namely Cars, buses, velocipedes motorcycles and trucks.

The augmentation in EV numbers brings a new issue that is the high electricity demand from the grid. One efficient solution to overcome the impact is to decentralize the power generation such as integrating renewable energy local sources into charging infrastructure. To address this challenge, Liu et al. [2] report the interaction between renewable energy and EVs charging problem in the presence of smart grid technologies.

PV power technology is expected to go through a substantial development in future, due to increased environmental awareness, the cost reduction and rise in efficiency of the PV modules. P. J. Tulpule et. al.[3] have cited several economic and environmental profits of the PV powered charging station in workplace parking. Furthermore, the charging operation is made during the daytime, which means the power generation is in its maximum point. Therefore a considerable cost saving is guaranteed. The installed PV modules on working parking garage's roof gives also free shelters in bad weather conditions[4]. Because of these advantages, the PV-grid based system is more preferred than other renewable energy based systems.

All electrically power assisted vehicles need to be recharged via charging systems, and the stations that use photovoltaic module as a source of electric energy for the battery recharging are called photovoltaic- charging stations (PVCS).

The PVCS is divided into two mandatory types, which are PV-grid charging system and PV-standalone charging systems. In this paper, we will investigate this topic by comparing the features of the two architectures and giving the actual technological status of charging system. For this reason, we present every part of the PV charging infrastructure to give an updated literature to engineers and researchers.

This paper is organized as follows: The second section gives an overview of the chargers standards.

The third section analyses the general architecture of the PVCS. In the fourth section, every component of the station is investigated

2. Electric Vehicle charging Standards

From a standardization point of view, there is three main organizations that work to standardize electrical characteristics of EV charging stations in the world: the Society of Automotive Engineering (SAE), CHAdeMO association and International Electrotechnical Commission (IEC). Besides these bodies, the worldwide leader of the electric carmakers, Tesla motors develops its own standards for its model S, Model X and Roadster electric cars.

Every organization cited above, offers a range of charger norms that work on both AC and another dedicated to DC voltage. In this paper we investigate only the data belong to DC range. For instance, the SAE has been working on standard J1772, which organises EV chargers into 3 main categories [5]: Level 1, Level 2 and Level 3. i) Level 1: the charger is on- board and provides DC voltage with maximum current of 80 A, and maximum power of 40 kW. ii) Level 2: the charger provides DC voltage with maximum current of 200 A, maximum power of 90 kW. iii) Level 3: charger is off- board. The charging station provides DC voltage directly to the battery via a DC connector, with a maximum power of 240 kW. All chargers from level 3 are considered as fast chargers. CHAdeMO and IEC proposed some power and current specifications concerning DC fast charging. For more information, a concise summary of power and current level evaluation for electric vehicle DC charging standards are listed in table 1.

Table 1. DC charging standards for EV.

Level	**Max current rating (A)**	**Max power rating (KW)**
SAE standard		
DC level 1	80	40
DC level 2	200	90

IOP Conf. Series: Materials Science and Engineering **353** (2018) 012014 doi:10.1088/1757-899X/353/1/012014

DC level 3	400	240
CHAdeMO		
DC fast charging	125	62,5
IEC Standard		
DC fast charging	400	100-200
Tesla Motor		
DC super-charger	340	136

3. Photovoltaic Charging station topologies

The PV-EV charging stations are divided into two categories, which are PV-grid charging system and PV-standalone charging systems. This section reviews the two architectures and gives a technical comparison between them.

3.1. PV-grid charging system

The literature on photovoltaic energy for EV charging is knowing an advanced and exponential development. This is due to the cost, sustainability and flexibility in integrating with the existing grid of the electricity supplied from PV modules.

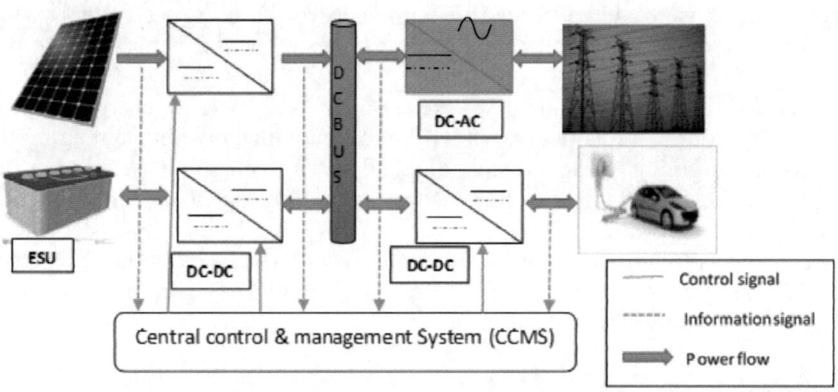

Figure 1. A bloc diagram of a PV-grid charging system.

The showed charging architecture in figure 1, which is studied from different points of view in various published paper[6], [7], is characterized by two conversion stages obtained through AC/DC and DC/DC converters. Furthermore, the dc bus has a high importance, because it is proposed to interface the PV array, the ESU and the EV battery pack combining other dc powered electronics. In this architecture, the batteries or energy storage unit (ESU) could be optional since the station is directly connected to the grid. Nevertheless, it would be a substantial part in case of willing reduce the dependence to the grid. In [8] the authors investigate the convivial topologies for integrated PV-grid charger.

3.2. Standalone PV charging system

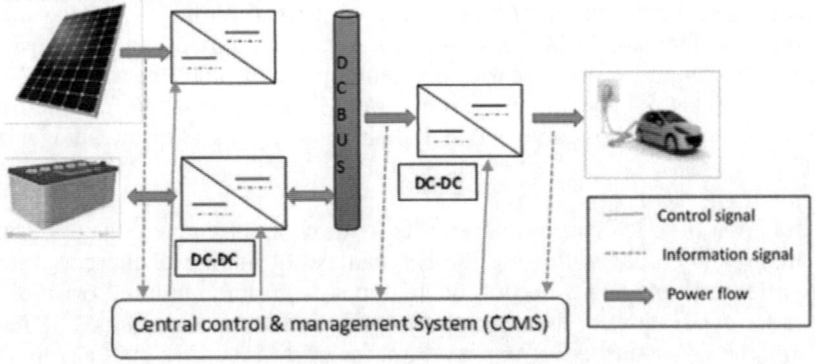

Figure 2. Standalone PVCS

In contrast with PV-EV charging station connected to grid, the standalone or off-grid station could provide energy to EV's batteries without any connection to the power grid. To this end, the charging system is necessarily equipped with an Energy storage unit (ESU) in order to be able to deliver continuously the power to the EV battery during night or when the PV modules cannot produce the sufficient energy[9]. For instance, In [10] authors proposed a standalone photovoltaic vehicle charge using second life lithium batteries as ESU.

4. Hardware infrastructure in PVCS: State of the art

4.1. PV system with MPPT control

To harvest electric energy from the sun light, PV modules are required to make the conversion. Numerous PV modules technologies exist in the market, namely polycrystalline and monocrystalline modules, thin film, heterogeneous intrinsic thin film[11], [12].The PV power nature is intermittent which depends strongly on weather conditions. Thus, the fluctuation in both solar irradiance and temperature results in nonlinear I–V and P–V characteristic curves, which makes the position of the Maximum Power Point (MPP) variable over time and difficult to be located and tracked. Because of this intermittent dynamic, the Maximum Power Point Tracker is crucial to harvest the maximum power through a dc-dc converter[13]–[15].

The basic principle of any MPPT command can be illustrated by Figure 3.We compare a power (P2) measured at time (t) with a power (P1) measured at time (t-1) : If the derivative is positive (P1 <P2), it means that we are in the left side of the MPP. In contrast, if the derivative is negative (P1 >P2), it means that we have exceeded the MPP. According to the actual condition, we increase or we decrease the duty cycle of the PWM control to match the optimal point.

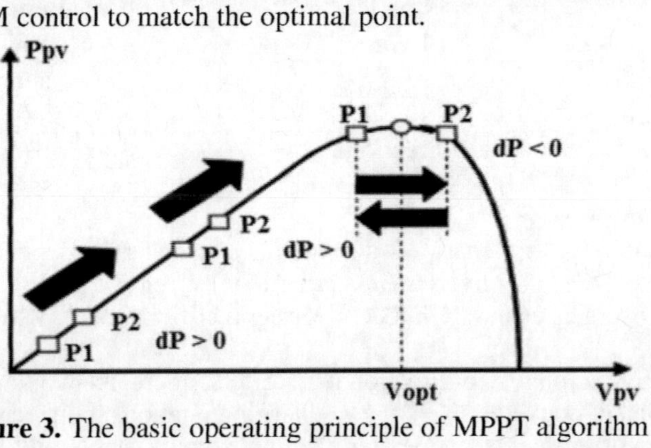

Figure 3. The basic operating principle of MPPT algorithm

In literature, various MPPT approaches have been developed which could devised to intelligent and classical strategies. In the first category we find: Fuzzy Logic, Neural Network, Extremum seeking control, Particle swarm optimization…[15]–[18] However, in the second category we find: Incremental Conductance Perturbation , Fractional Open-Circuit Voltage, and the most widely used technique is Perturb and Observe (P&O) algorithms due to its simplicity and fast response [13], [14], [19]. In addition, a comparison between conventional and intelligent MPPT techniques are investigated in [20], [21]. For more investigation, some interesting papers review nearly all known MPPT techniques and give a comparative between their advantages and disadvantages are in [22]–[24]

4.2. Bidirectional DC Charger

The mandatory function of a dc charger in PVCS is to allow an effective control during the charging process by interfacing the dc bus voltage to the EV battery. In literature, charger systems for electric vehicles are classified into off-board and on-board types with unidirectional or bi-directional power flow[25]. This latter type adopts the Vehicle to grid (V2G) concept, since it has not only the capability to charge the EV in one effective way from the grid or the ESU, but also it supports also the power flow in the other direction back to the grid[26]–[28].

Hence, it gives the grid the opportunity to benefit from the energy stored in the EV battery during shortage energy time. Beside this, V2G technology require advanced communication components to guarantee a safe operation and a smart grid to accept the power injection[2], [29].The impact of the V2G on the EV battery pack is evaluated and discussed more deeply in[28], [30].

Concerning the power converter configuration used, we find essentially two structures, namely a non-isolated and isolated converters. This latter type ensures the galvanic isolation, by contrast, it presents some drawbacks such as high cost realization due to adding transformers. In the other side, the non-isolated bidirectional dc chargers are considered the most appropriate to serve as DC chargers because of their compactness and higher reliability[31].

In [33] authors reviewed the topologies of non-isolated bidirectional DC-DC chargers and proposed application of a rapid charging station at municipal parking decks. In [34] an on-board bidirectional soft-switched battery charger is proposed. Serving the same purpose, The Interleaved design of Bidirectional converter and a half bridge topology are studied in [35], [36] with the aim to minimize the inductor size and increase the efficiency using soft-switching control.

On-board chargers are usually used to increase the charge availability for all kind of rechargeable vehicles. However, some constraints must be taken in consideration such as weight, size and power rate since this type is usually used for slow charging during nights or embedded on the vehicle. Therefore there will be an outlook for PV charging stations to move from slow onboard chargers to high-power off-board fast chargers in the future. The dissimilarities between the On-board and OFF-board Dc charger are summarized in the Table 2.

Table 2. Comparison between on/off board dc chargers

	On-board DC charger	**Off Board DC charger**
Size	small	medium or big
Wight	light	heavy
Charging time	long	short
Power range	< 40 KW	< 240 KW
Power flow	Bidirectional	Mono / Bidirectional

4.3. Bidirectional inverter

The bidirectional inverter for EV charging system play a dual role depending on the current flow direction. It functions as DC-AC converter (inverter mode) when the grid requires the power from the dc bus. In contrast, it operates as an AC-DC converter (rectifier mode), when the DC link need to be supplied from the grid.

Among different configurations of bidirectional inverters, there is two famous categories namely, isolated and non-isolated converters. For PV charging stations connected to grid, the isolated configuration is the favorite one as it operates in high voltage levels and could provide a galvanic isolation between Alternative current part and direct current part of the system .[37], [38].

The key constrained point of the inverter design is the frequency stabilization. Hence, a good converter must have the capability to converter voltage signal nature with suitable synchronization frequency with the power grid and conserve the grid power quality.

4.4. Energy storage system

Due to the intermittent nature of the generated photovoltaic power, the energy storage system (ESS) is indispensable in the charging station, especially for the isolated ones. The role of this ESS is to compensate or absorb the difference between the generated photovoltaic power and the required load power[4] in reduce power requirements from main grid during charging operations [39] For this reason, it is considered as the key element of the energy management in any renewable sources based system [40]. A comprehensive analysis of the ESS concerning different power storage techniques, their efficiencies and economic evaluation is reported in [40].

In general, lead-acid batteries are the most popular storage technology used due to their reduced cost and long lifespan. However, their energy density is small and to offer the sufficient power amount, the lead pack takes a huge volume. With the inauguration of new huge factories deduced to manufacture only lithium ion batteries, the cost becomes cheaper as well as the lead technology. Furthermore, in order to alleviate the impact of dynamic power interactions on battery's lifetime, hybrid energy storage unit (HESS) is proposed. This hybrid ESU may contains, in addition to batteries, supercapacitors, flywheels[41], [42].

4.5. Energy management system

There are many papers that investigate the power management in EV charging system, In the study [45] authors proposed an intelligent energy management system with wireless ZigBee communication. An energy management strategy (EMS) using an artificial neural network is described in[46] in order to shave the domestic peak grid load by the coordinated response of distributed energy resource. Additionally, an offline particle swarm optimization (PSO) is performed besides a dynamic programming (DP) as an online reactive management layer in the study research [47], with the aim to continuously minimize the power flow operation cost. Additionally, a consistent communication is required between the charging management system and the battery management system is usually maintained via the CAN in order to ensure the security and the integrity.

5. Conclusion

This paper briefly reviews works recently conducted in the area of photovoltaic DC charging stations for EVs. A discussion and a comparative study between different components of the charging stations is done.

To summarize, the benefits of PV-EV chargers that adopt V2G technology will receive an increased attention and more investments from grid operators and carmakers in the future.As conclusion, The Photovoltaic charging structure is becoming more complex with several functions integrated into the system, which require intelligent controls in each block and real time management for the whole station.

References

[1] C. Samaras and K. Meisterling, 'Life Cycle Assessment of Greenhouse Gas Emissions from Plug-in Hybrid Vehicles: Implications for Policy', Environ. Sci. Technol., vol. 42, no. 9, pp. 3170–3176, May 2008.

[2] L. Liu, F. Kong, X. Liu, Y. Peng, and Q. Wang, 'A review on electric vehicles interacting with renewable energy in smart grid', Renew. Sustain. Energy Rev., vol. 51, pp. 648–661, 2015.

[3] P. J. Tulpule, V. Marano, S. Yurkovich, and G. Rizzoni, 'Economic and environmental impacts of a PV powered workplace parking garage charging station', Appl. Energy, vol. 108, pp. 323–332, 2013.

[4] R. H. Ashique, Z. Salam, M. J. B. A. Aziz, and A. R. Bhatti, 'Integrated photovoltaic-grid dc fast charging system for electric vehicle: A review of the architecture and control', Renew. Sustain. Energy Rev., 2016.
[5] L. Dickerman and J. Harrison, 'A New Car, a New Grid', IEEE Power Energy Mag., vol. 8, no. 2, pp. 55–61, Mar. 2010.
[6] H. Hõimoja, A. Rufer, G. Dziechciaruk, and A. Vezzini, 'An ultrafast EV charging station demonstrator', in Power Electronics, Electrical Drives, Automation and Motion (SPEEDAM), 2012 International Symposium on, 2012, pp. 1390–1395.
[7] S. Bai, D. Yu, and S. Lukic, 'Optimum design of an EV/PHEV charging station with DC bus and storage system', in Energy Conversion Congress and Exposition (ECCE), 2010 IEEE, 2010, pp. 1178–1184.
[8] N. Naghizadeh and S. S. Williamson, 'A comprehensive review of power electronic converter topologies to integrate photovoltaics (PV), AC grid, and electric vehicles', in 2013 IEEE Transportation Electrification Conference and Expo (ITEC), 2013, pp. 1–6.
[9] O. Ekren and B. Y. Ekren, 'Size optimization of a PV/wind hybrid energy conversion system with battery storage using simulated annealing', Appl. Energy, vol. 87, no. 2, pp. 592–598, 2010.
[10] S. J. Tong, A. Same, M. A. Kootstra, and J. W. Park, 'Off-grid photovoltaic vehicle charge using second life lithium batteries: An experimental and numerical investigation', Appl. Energy, vol. 104, no. Supplement C, pp. 740–750, Apr. 2013.
[11] O. Vetterl et al., 'Intrinsic microcrystalline silicon: A new material for photovoltaics', Sol. Energy Mater. Sol. Cells, vol. 62, no. 1, pp. 97–108, 2000.
[12] A. H. Fanney, M. W. Davis, and B. P. Dougherty, 'Short-term characterization of building integrated photovoltaic panels', in ASME Solar 2002: International Solar Energy Conference, 2002, pp. 211–221.
[13] S. Li, H. Liao, H. Yuan, Q. Ai, and K. Chen, 'A MPPT strategy with variable weather parameters through analyzing the effect of the DC/DC converter to the MPP of PV system', Sol. Energy, vol. 144, no. Supplement C, pp. 175–184, Mar. 2017.
[14] J. Ahmed and Z. Salam, 'An improved perturb and observe (P&O) maximum power point tracking (MPPT) algorithm for higher efficiency', Appl. Energy, vol. 150, no. Supplement C, pp. 97–108, Jul. 2015.
[15] S. L. Brunton, C. W. Rowley, S. R. Kulkarni, and C. Clarkson, 'Maximum Power Point Tracking for Photovoltaic Optimization Using Ripple-Based Extremum Seeking Control', IEEE Trans. Power Electron., vol. 25, no. 10, pp. 2531–2540, Oct. 2010.
[16] P. S. Gavhane, S. Krishnamurthy, R. Dixit, J. P. Ram, and N. Rajasekar, 'EL-PSO based MPPT for Solar PV under Partial Shaded Condition', Energy Procedia, vol. 117, no. Supplement C, pp. 1047–1053, Jun. 2017.
[17] C. S. Chiu, 'T-S Fuzzy Maximum Power Point Tracking Control of Solar Power Generation Systems', IEEE Trans. Energy Convers., vol. 25, no. 4, pp. 1123–1132, Dec. 2010.
[18] M. A. Sahnoun, H. M. R. Ugalde, J.-C. Carmona, and J. Gomand, 'Maximum Power point Tracking Using P&O Control Optimized by a Neural Network Approach: A Good Compromise between Accuracy and Complexity', Energy Procedia, vol. 42, no. Supplement C, pp. 650–659, Jan. 2013.
[19] D. Sera, L. Mathe, T. Kerekes, S. V. Spataru, and R. Teodorescu, 'On the Perturb-and-Observe and Incremental Conductance MPPT Methods for PV Systems', IEEE J. Photovolt., vol. 3, no. 3, pp. 1070–1078, Jul. 2013.
[20] H. Bounechba, A. Bouzid, K. Nabti, and H. Benalla, 'Comparison of Perturb & Observe and Fuzzy Logic in Maximum Power Point Tracker for PV Systems', Energy Procedia, vol. 50, no. Supplement C, pp. 677–684, Jan. 2014.
[21] K. Sundareswaran, V. Vignesh kumar, and S. Palani, 'Application of a combined particle swarm optimization and perturb and observe method for MPPT in PV systems under

partial shading conditions', Renew. Energy, vol. 75, no. Supplement C, pp. 308–317, Mar. 2015.

[22] M. A. Danandeh and S. M. Mousavi G., 'Comparative and comprehensive review of maximum power point tracking methods for PV cells', Renew. Sustain. Energy Rev.

[23] F. T. Sawant, P. L. Bhattar, and C. L. Bhattar, 'Review on maximization of solar system under uniform and non uniform conditions', in 2017 International Conference on Circuit ,Power and Computing Technologies (ICCPCT), 2017, pp. 1–7.

[24] A. Khare and S. Rangnekar, 'A review of particle swarm optimization and its applications in Solar Photovoltaic system', Appl. Soft Comput., vol. 13, no. 5, pp. 2997–3006, May 2013.

[25] M. Yilmaz and P. T. Krein, 'Review of Battery Charger Topologies, Charging Power Levels, and Infrastructure for Plug-In Electric and Hybrid Vehicles', IEEE Trans. Power Electron., vol. 28, no. 5, pp. 2151–2169, May 2013.

[26] G. R. C. Mouli, P. Bauer, and M. Zeman, 'Comparison of system architecture and converter topology for a solar powered electric vehicle charging station', in 2015 9th International Conference on Power Electronics and ECCE Asia (ICPE-ECCE Asia), 2015, pp. 1908–1915.

[27] J. Channegowda, V. K. Pathipati, and S. S. Williamson, 'Comprehensive review and comparison of DC fast charging converter topologies: Improving electric vehicle plug-to-wheels efficiency', in 2015 IEEE 24th International Symposium on Industrial Electronics (ISIE), 2015, pp. 263–268.

[28] M. Yilmaz and P. T. Krein, 'Review of the Impact of Vehicle-to-Grid Technologies on Distribution Systems and Utility Interfaces', IEEE Trans. Power Electron., vol. 28, no. 12, pp. 5673–5689, Dec. 2013.

[29] M. G. Vayá and G. Andersson, 'Integrating renewable energy forecast uncertainty in smart-charging approaches for plug-in electric vehicles', in PowerTech (POWERTECH), 2013 IEEE Grenoble, 2013, pp. 1–6.

[30] J. D. Bishop, C. J. Axon, D. Bonilla, M. Tran, D. Banister, and M. D. McCulloch, 'Evaluating the impact of V2G services on the degradation of batteries in PHEV and EV', Appl. Energy, vol. 111, pp. 206–218, 2013.

[31] S. Chakraborty, T. Senjyu, H. Toyama, A. Y. Saber, and T. Funabashi, 'Determination methodology for optimising the energy storage size for power system', IET Gener. Transm. Distrib., vol. 3, no. 11, pp. 987–999, 2009.

[32] M. Vaidya, E. K. Stefanakos, B. Krakow, H. C. Lamb, T. Arbogast, and T. Smith, 'Direct DC-DC electric vehicle charging with a grid connected photovoltaic system', in Conference Record of the Twenty Fifth IEEE Photovoltaic Specialists Conference - 1996, 1996, pp. 1505–1508.

[33] Y. Du, X. Zhou, S. Bai, S. Lukic, and A. Huang, 'Review of non-isolated bi-directional DC-DC converters for plug-in hybrid electric vehicle charge station application at municipal parking decks', in Applied Power Electronics Conference and Exposition (APEC), 2010 Twenty-Fifth Annual IEEE, 2010, pp. 1145–1151.

[34] S. Ebrahimi, R. Akbari, F. Tahami, and H. Oraee, 'An Isolated Bidirectional Integrated Plug-in Hybrid Electric Vehicle Battery Charger with Resonant Converters', Electr. Power Compon. Syst., vol. 44, no. 12, pp. 1371–1383, Jul. 2016.

[35] A. K. Rathore and U. R. Prasanna, 'Analysis, design, and experimental results of novel snubberless bidirectional naturally clamped ZCS/ZVS current-fed half-bridge DC/DC converter for fuel cell vehicles', IEEE Trans. Ind. Electron., vol. 60, no. 10, pp. 4482–4491, 2013.

[36] S. Dusmez, A. Khaligh, and A. Hasanzadeh, 'A zero-voltage-transition bidirectional dc/dc converter', IEEE Trans. Ind. Electron., vol. 62, no. 5, pp. 3152–3162, 2015.

[37] Y. Xia, W. Wei, Y. Peng, P. Yang, and M. Yu, 'Decentralized Coordination Control for Parallel Bidirectional Power Converters in a Grid-Connected DC Microgrid', IEEE Trans. Smart Grid, vol. PP, no. 99, pp. 1–1, 2017.

[38] S. Kouro, J. I. Leon, D. Vinnikov, and L. G. Franquelo, 'Grid-Connected Photovoltaic Systems: An Overview of Recent Research and Emerging PV Converter Technology', IEEE Ind. Electron. Mag., vol. 9, no. 1, pp. 47–61, Mar. 2015.

[39] C. Capasso and O. Veneri, 'Experimental study of a DC charging station for full electric and plug in hybrid vehicles', Appl. Energy, vol. 152, pp. 131–142, Aug. 2015.

[40] H. Ibrahim, A. Ilinca, and J. Perron, 'Energy storage systems—characteristics and comparisons', Renew. Sustain. Energy Rev., vol. 12, no. 5, pp. 1221–1250, 2008.

[41] G. Joos, M. de Freige, and M. Dubois, 'Design and simulation of a fast charging station for PHEV/EV batteries', in 2010 IEEE Electrical Power Energy Conference, 2010, pp. 1–5.

[42] C. Spataru, Y. C. Kok, and M. Barrett, 'Physical energy storage employed worldwide', Energy Procedia, vol. 62, pp. 452–461, 2014.

[43] W. Jing, C. H. Lai, S. H. W. Wong, and M. L. D. Wong, 'Battery-supercapacitor hybrid energy storage system in standalone DC microgrids: areview', IET Renew. Power Gener., vol. 11, no. 4, pp. 461–469, 2017.

[44] M. d R. A. Calado, S. J. P. S. Mariano, J. A. N. Pombo, and R. J. C. Pinto, 'PV charging station for electric vehicles: Management and interface system', in 2016 IEEE 16th International Conference on Environment and Electrical Engineering (EEEIC), 2016, pp. 1–5.

[45] P. Kulshrestha, L. Wang, M. Y. Chow, and S. Lukic, 'Intelligent energy management system simulator for PHEVs at municipal parking deck in a smart grid environment', in 2009 IEEE Power Energy Society General Meeting, 2009, pp. 1–6.

[46] K. Mahmud, S. Morsalin, M. J. Hossain, and G. E. Town, 'Domestic peak-load management including vehicle-to-grid and battery storage unit using an artificial neural network', in 2017 IEEE International Conference on Industrial Technology (ICIT), 2017, pp. 586–591.

[47] M. O. Badawy and Y. Sozer, 'Power Flow Management of a Grid Tied PV-Battery System for Electric Vehicles Charging', IEEE Trans. Ind. Appl., vol. 53, no. 2, pp. 1347–1357, Mar. 2017.

Mechanical characterization of new micro-composites composed by natural clay matrix and PEG 6000 fillers

Mostapha El Jai[1], Iatimad Akhrif[2,4], Laila Mesrar[3], Raouf Jabrane[4]

[1]Euromed School of Civil Engineering, Euromediterranean University of Fez, Morocco
[2] Laboratoire Systèmes et Environnements Durables (LSED), Faculté des Sciences de l'Ingénieur (FSI), Université Privée de Fès (UPF), Fez, Morocco
[3]LOMC, Le Havre University, France
[4]Geosciences and Environment Laboratory, Science and Technique Faculty, Sidi Mohamed Ben Abdellah University, Fez, Morocco

m.eljai@ueuromed.org

Abstract. The aim of this paper is to characterize mechanically the new micro-composites that have been developed in our laboratories. The composites are composed by natural clay (as a matrix) with variant percentages of Polyethylene Glycol 6000 (PEG 6000) as micro-fillers. We used the compression test for the measurement of the static parameters such as elasticity modulus in elastic region and the hardening coefficient which permits to describe the plasticity behaviour of the materials. An additional energetic approach is proposed in order to quantify the evolution of the plasticity of the reinforced materials, caused by the PEG 6000, for different percentages of this polymer.

Key-words: Microcomposite/clay matrix/PolyEthylene Glycol 6000, Mechanical properties, Elasticity, Plasticity, energetic analysis.

1. Introduction

This work is the third step of a large study that have been started by the elaboration of new micro-composites made by clay matrix and the addition of the PEG 6000, in order to valorize the mechanical, geotechnical and thermal properties of the original clay material of Fez region, Morocco. In a previous works (Akhrif & al. 2014, 2015 and 2016), we have presented and discussed the elaboration protocols of the materials, the microstructural transformations according to XRD spectra, and thermal analysis according to the DSC characterization.

The aim of this article is to quantify the mechanical behavior according to compression tests. The characterization was performed on cylinders that are presented in the next sections. The findings correspond to Young modulus, the mechanical strength and the loss of mechanical energy due to plastification of the materials.

2. Material and methods

2.1. Materials

2.1.1. Natural clay used

IOP Conf. Series: Materials Science and Engineering **353** (2018) 012015 doi:10.1088/1757-899X/353/1/012015

The clay matrix used corresponds to natural marl/clay from Ben Jellik quarry localized in the east of Fez city as shown in figure 1. Figure 1 shows also the Lambert coordinates of the region that correspond to "Longitude: 34°02'14.784" – Latitude: -4°56'56.642" – Altitude: 431 m". A lithographic section, given in figure 2 permits to see the geologic formation of the extraction region.

The clay used corresponds to the Miocene marl, shown on figure 2 by a red star, extracted on a depth of 20 m. According to XRD analysis [1] [4], the material is composed by the clay phases illite and Kaolinite, calcite and quartz essentially.

Fig. 1. Localization of the extraction site (Longitude: 34°02'14.784" – Latitude: -4°56'56.642" – Altitude: 431 m)

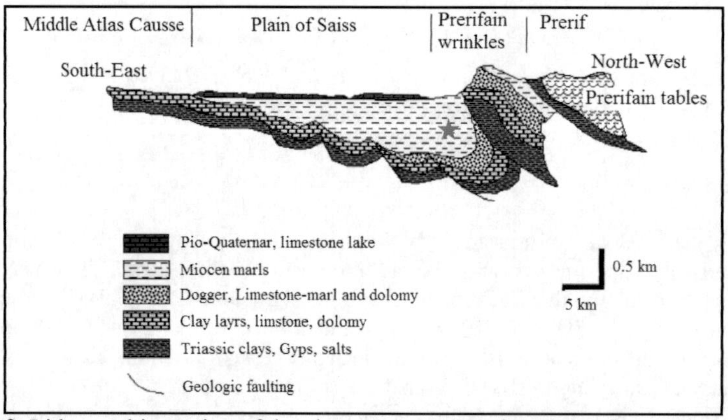

Fig. 2. Lithographic section of the clay source: extraction shown by a red star [1]

2.1.2. Polyethylene Glycole 6000

The polyethylene Glycole 6000 belongs to the polyethylene oxide class with a molar mass of 6000 g/mole. This is the simplest of the hydro-soluble polymer that exists [2]. The PEG 6000 is ecologic, biodegradable and biocompatible polymer [3]. The chemical structure is $-(CH2-CH_2-O-)_n$. this material is semi-crystalline having a crystallinity ratio between 75% and 85% [2].

2.2. Methods

2.2.1. Elaboration of the materials

The microcomposites developed were obtained by a succession of grindings from the initial quarry rock, to the fine elements. The PEG 6000 has been included with the finest clay within the electrical Retsch RM 200 mortar [5] for 15 minute of simultaneous mixing and grinding.

Fig. 3. The grinder used to mixing Clay matrix and PEG 6000 [4]

Thereafter, the compression cylinders were processed using a metallic mold. The initial diameter of each cylinder was about 25 mm, and the initial height was about 30 mm. Shrinking phenomenon was described and modeled in [4] and [6]. The cylinders were after polished and destroyed in the compression test machine.

2.2.2. Material processing parameters

The materials processing corresponds to a complete experiments plan, according to:

- 3 levels of grinding sizing, denoted 3, 5 and 7 (the finest size), related to the grinder scales;
- 6 levels of PEG 6000 percentages per 6g of clay material: 0.1g, 0.2g, to 0.6g.

Thus, the total of the combinations corresponds to 18, in addition with the compression test applied to the row materials (clay) according to the 3 different sizing. To ensure the repetitiveness of the test, we adopt 2 samples by each composition. The total compression tests correspond to:

$$N = (6+1) \times 3 \times 2 = 42 \text{ compression tests}$$

2.2.3. Compression test procedure

The compression tests were performed using "EM 550-DeltaLab machine" with a maximal strength of 50 kN. The compression parameters are:

- Static compression, where the superior plate speed is 0.2 mm/s; the inferior plate is blocked;
- Force and displacement acquisition time step: 0.5 s;
- Ambient temperature.

3. Results

3.1. Mechanical properties

The first findings correspond to the mechanical properties: Young modulus "E", mechanical strength "Rm" and plasticity coefficient known as hardening coefficient "n", according to the follow models [7]:

- Elastic domain modelled by Hook law :

$$\sigma = E\,\varepsilon \quad (1)$$

- Plastic domain modelled by Ludwig law :

$$\sigma = \sigma_0 + k\,\varepsilon^n \quad (2)$$

3.1.1. Mechanical strength

The mechanical strength corresponds to the maximum of stress σ. Figure 4 presents the distribution of the mechanical strength of the materials according to the two parameters: grinding size and the polymer concentrations. We observe that there is no significant evolution of Rm with the two parameters. The values of Rm oscillate around its initial values corresponding to the row clay material related to 0% of PEG.

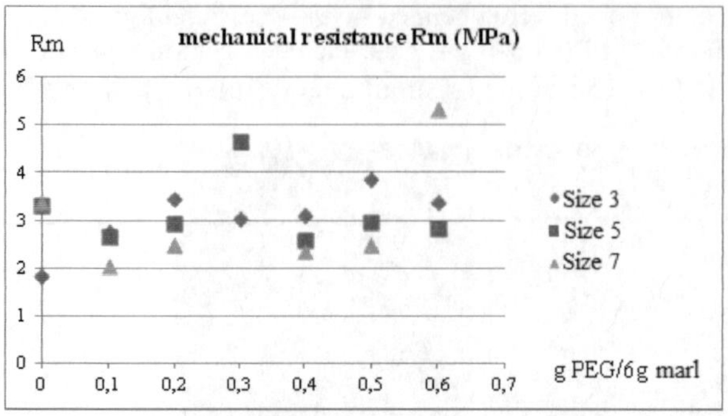

Fig. 4. Mechanical strength of the materials according to grinding size and PEG 6000 concentration

3.1.2. Young modulus

On figure 5-a, we can see that the Young modulus oscillates around an average value, and there is no significant evolution according to the evolution parameters. For each grinding size, we fit a regression line. The linear regression coefficient R^2 calculated is very low, so the linear modelling is note accepted. Figure 5-b plot the average of the Young modulus according to clays size; thus, the curve gives the behaviour of E(MPa) according to PEG 6000 concentration.

a) 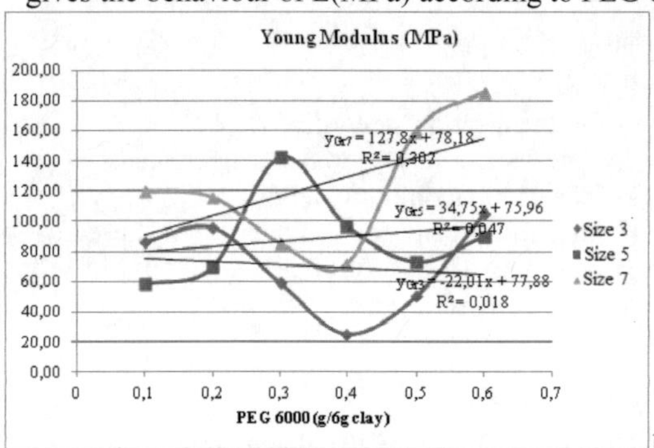

b) 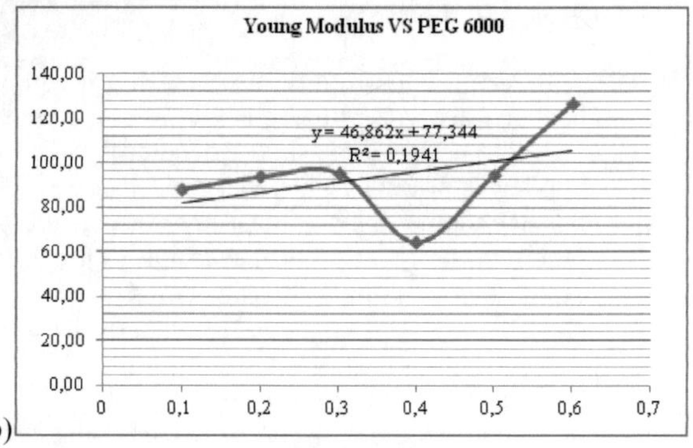

Fig. 5. Young Modulus according to a) grinding size and PEG 6000 b) PEG 6000

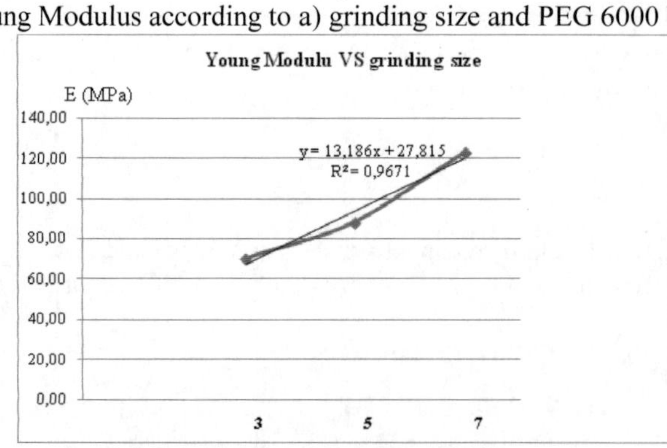

Fig. 6. Young Modulus VS Grinding scale

Figure 6 shows that the Young modulus increases significantly with the finenesses of grinding. This behavior is realistic since the finenesses of grinding permits a best agglomeration of clay material in general.

3.1.3. Hardening coefficient n

The figures 7-a) and b) show the evolution of the hardening coefficient "n" according to the PEG 6000 and grinding size. We observe a great decrease of this coefficient according to PEG 6000.

According to figure 7-a), we can say that the material evaluate from a perfectly elastic material for the configuration (size 3, PEG 0.1 g) with a value of "n=1" to a perfectly plastic material for the configuration (size 7, PEG 0.6 g) having a value of "n=0.12". Figure 7-a) shows the regression line with a R^2 up to 0.7. Thus, the plastification of the material follows a linear law according to the PEG 6000 addition.

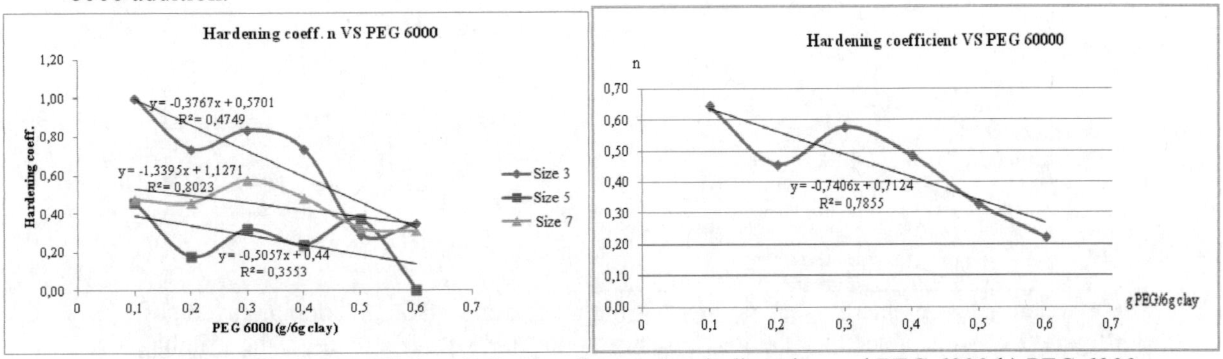

Fig. 7. Hardening coefficient "n" according to a) grinding size and PEG 6000 b) PEG 6000

3.1.4. Energetic approach

The energetic analysis permits us to quantify the significant evolution of the materials according to the addition of the polymer. The approach is to evaluate the loss of energy caused by plastification phase. So for each material:

- According to the values of the Young modulus, we suppose a perfectly elastic material. We estimate the theoretical elastic energy that has to be created within solicitation;
- We calculate in parallel the value of plastic energy transmitted within the material, according to the result of compression test
- The difference between the two energy quantities corresponds to the loss of energy caused by plastification.
- Thereafter, we estimate the loss of energy related to the theoretical elastic energy in term of percentage, as shown in equation 3 below :

$$Relative\ loss\ of\ energy\ T(\%) = \frac{Loss}{theoretical\ Elastic\ energy} \quad (3)$$

Figures 8-a), b) and c) show the evolution of the energetic ratio T according to the grinding size and PEG 6000 concentrations. We can observe a really significant and linear evolution of the relative loss of energy according to the PEG concentration. The figures 9-a) and b) shows the average of the energetic indicator T according to the parameters (size and PEG) separately.

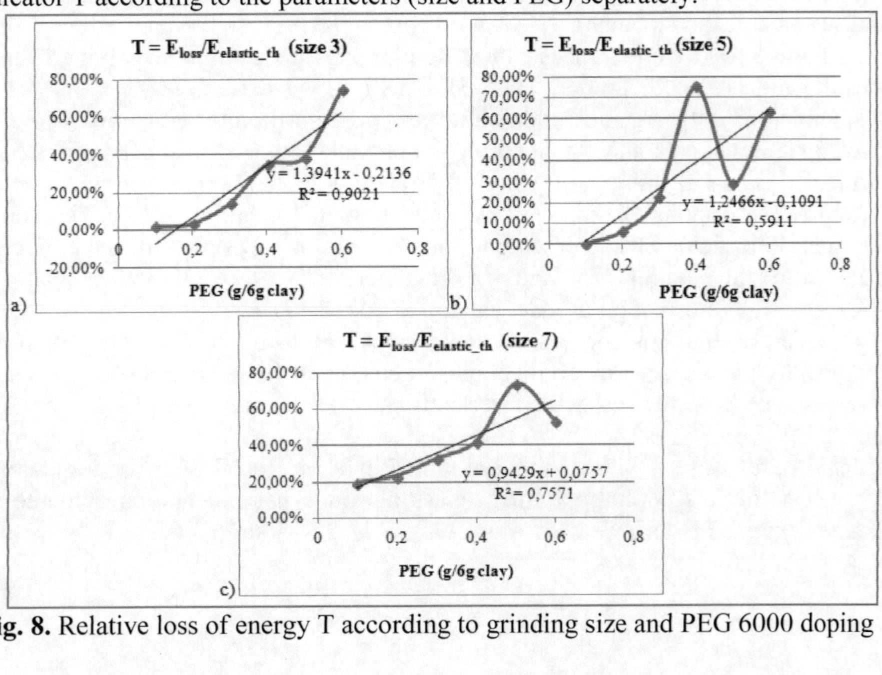

Fig. 8. Relative loss of energy T according to grinding size and PEG 6000 doping

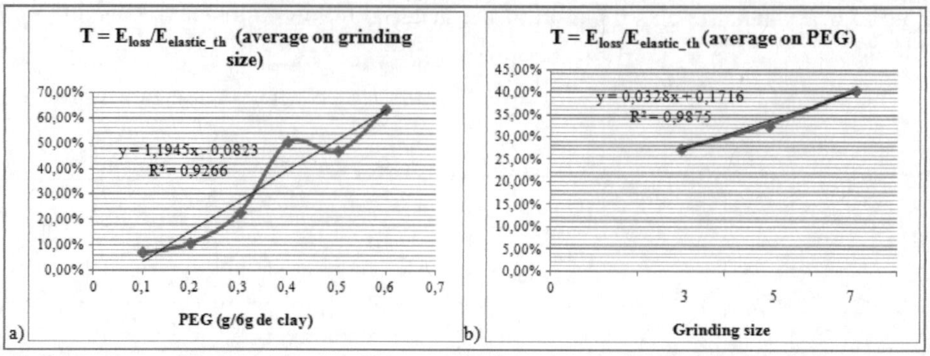

Fig. 9. Relative loss of energy T according to grinding size and PEG 6000 doping (averages)

From figure 9-b, we can see that the loss of energy increase also according to the grinding size. From our point of view, we interpret this result by the increasing of the inter-particle water that could exist within the clay matrix with the increasing of the finenesses of the grinding.

4. Conclusion

The results presented in this work permit us to see that a simple procedure of a dry controlled grinding permits a significant evolution of plastification of the natural and not purified clay with the addition of the PEG 6000. Indeed the addition of the polymer does not affect the mechanical properties in the elastic domain, but in the plastic domain, the effect is very significant: the clay material passed from a perfectly elastic material to a quasi perfectly plastic material in the plastic domain.

A future step consists on the performing of more mechanical essays, related to vibration behavior of the new materials, to estimate the possibility of existing of a hysteresis related to charging/discharging cycles. If existing, the new materials or even the PEG polymer could be used in reinforcing ground against vibration and seismic solicitations.

References

[1] Mesrar Laila (2013), Maryam Akdim, Mohamed Lakrim, Omar El Aroussi, Iatimad Akhrif, Abdel-Ali Chaouni & Raouf Jabrane, « Echnological valorization of the miocene clay in the region of fez morocco): characterisation and exploitation possibilities », Present environment and sustainable development, vol. 7, no. 1, pp. 310-317.

[2] Hassouna Fatima, « Etude des mécanismes de phototransformation de polymères hydrosolubles en milieu aqueux », PhD report, BLAISE PASCAL University (2006).

[3] Sarier & Onder, (2010) Organic Modification of Montmorillonite with Low Molecular Weight Polyethylene Glycols and Its Use in Polyurethane Nanocomposite Foams. Thermochimica Acta , 510, 113-121. http://dx.doi.org/10.1016/j.tca.2010.07.004.

[4] Akhrif Iatimad, Mostapha El Jai, Laila Mesrar, Ahmed Elkhalfi, Abdelhamid Touache, Raouf Jabrane, "Physical Characterization and Elaboration Discussion of a Clay-PEG 6000 Composite with Natural Clay Matrix », Engineering, 2014, 6, 338-354.

[5] RETSCH RM 200, Information product catalog, 07/04/2017.

[6] Akhrif I., « Valorisation des marnes miocènes de la régions de Fès par élaboration de matériaux composites à matrice marne/argile et renfort en polyéthylène Glycole 6000 », Geosciencde & Environment Laboratory, Faculty of Sciences & Techniques, Sidi Mohamed Ben Abdellah University, Fez, Morocco.

[7] Collin Jean-Marc, Gérard Mauvoisin, Philippe Pilvin et Rochdi El Abdi, « Identification des paramètres de lois de comportement élastoplastique par indentation continue sphérique », 18ème Congrès Français de Mécanique Grenoble, 27-31 août 2007.

IOP Conf. Series: Materials Science and Engineering **353** (2018) 012016 doi:10.1088/1757-899X/353/1/012016

Securing Digital Images Integrity using Artificial Neural Networks

Tarik HAJJI, Zakaria ITAHRIOUAN & Mohammed OUAZZANI JAMIL
Laboratoire Systèmes et Environnements Durables (LSED), Faculté des Sciences de l'Ingénieur (FSI), Université Privée de Fès (UPF) , Fez, Morocco

hajji@upf.ac.ma

Abstract. Digital image signature is a technique used to protect the image integrity. The application of this technique can serve several areas of imaging applied to smart cities. The objective of this work is to propose two methods to protect digital image integrity. We present a description of two approaches using artificial neural networks (ANN) to digitally sign an image. The first one is "Direct Signature without learning" and the second is "Direct Signature with learning". This paper presents the theory of proposed approaches and an experimental study to test their effectiveness.

1. Introduction

Watermarking Is a technique to add copyright information or other verification information to a file audio, video, image or any other document. The message added to initial information, usually called brand, is a set of bits whose content depends on the application. The brand can be the name or identifier of the creator, owner, buyer or a form of signature that describes the host signal [1].
Artificial Neural Networks (ANN) is a problem resolution mechanism, inspired by the human brain reasoning. The ANN basic unit is the artificial neuron. An ANN is made up of a neurons collections organized in layers: The first layer the input layer, the last is the output layer and the layers in the middle are called the hidden layers 1 (Figure 1).

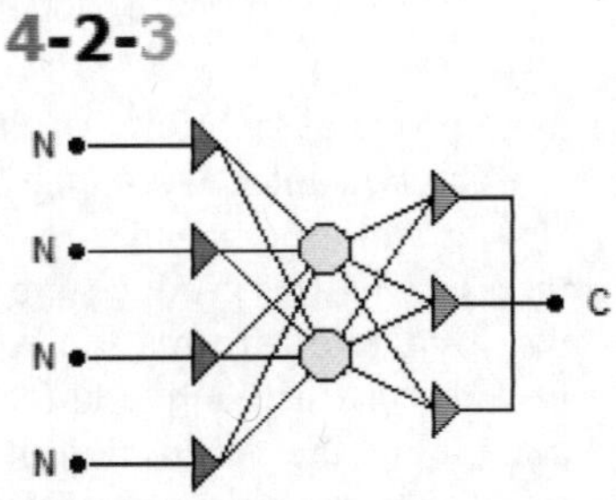

Figure 1. Artificial Neural Networks

An ANN is characterized by a specific architecture to define its neural structure. In this work we propose a notation to define this architecture, for example the notation $ANN\,(64, 16, 64)$ indicates the use of a multilayer ANN compound by:

- An input layer composed by 64 artificial neurons,
- One hidden layer composed by 16 artificial neurons,
- An output layer composed by 64 artificial neurons.

We distinguish two ways of using ANN:

- Use with learning: We have a database composed by a partial solution of the problem. Then, we use an algorithm to teach ANN the resolving mechanism of the different instances of the problem. The gradient back propagation is a famous algorithm able to comprehend an ANN the comprehension mechanisms. Finally, we can use the ANN after learning to solve new instances of the problem.

- Use without learning: In some cases, we have no information about the solution space for the problem studied. For example, the problem of classification which we do not have classes to classify elements. So we can use an ANN in unsupervised mode to find the classes firstly (clustering).

This paper presents in the first part some performed researches in the area of digital image signature. The second part contains the methodology used to sign digital images presented by our approach. The last part is dedicated to results presentation and discussion.

2. Digital signature

2.1. *State of the art*

Before developing our approach of digital image signature, we started by identifying research work in the same field. Baltzakis & al. presents a new online system of signature recognition and verification. The proposed system is based on texture features. In the first step, the classifier combines the decision results between the neural network and the Euclidean distance obtained by using the three sets of features. The results of the classifier of the first stage are used as input of the function of the second step (radial base structure of the ANN) which makes the final decision. The entire system has been extensively tested and resulted in a strong recognition and verification rate [2].

Sabourin & al. propose an ANN approach to build the first stage of an automatic system of handwritten signature verification. The directional probability density function was used as an overall form factor and power of discrimination has been strengthened by reducing its cardinal by filtering. Different experimental protocols were used to implement the ANN.As a comparison, on the same database and the same decision rule, we show that the BPN classifier is significantly better than the classifier threshold and it is compared favourably with the classifier k-nearest neighbour [3].

Yin Ooi & al. propose the hybrid methods of discrete Radon transform, principal component analysis, and probabilistic ANN. The proposed framework is designed to distinguish genuine signatures counterfeit depending on the image level. They realised experiments on their own database and on the MYCT database. Equal error rate (EER) of 1.51%, 3.23% and 13.07% are reported, respectively, for false random, casual and skilled in their own database. By working on the basis of MCYT signatures, their proposed approach strikes an AIR of 9.87% with 10 training samples [4].

2.2. *signature verifiers*

The objective of signature verification is to prove authenticity of the image form a regular basis. The signature verification techniques used are divided into two categories: online (dynamic) and offline (static).

The online approach uses a compressed tablet produced by digital compression to extract information about the signature and takes dynamic information such as compression speed, writing speed, number of stroke order and pen pressure at each point [5]. The application fields of this technique include the protection of the computers, the permission of computer users to access sensitive data or programs and people authentication for access to physical devices or buildings [6].

The offline approach involves less electronic control and uses signature images captured by camera or scanner [7]. An offline signature verification system uses features extracted from the digitized signature image. The characteristics used for online signature verification are simple and invariant [5].therefore, the only pixel image must be estimated. However, the offline systems are difficult to design, because many of the characteristics are recognized immediately [6].

3. Methodology

3.1. *Direct signature without learning*

We started our work by simply using ANN to sign a digital image. Two ANN was used without learning: (64, 4) and (64, 16, 4). After a random initialization of ANN synapses, we generated the signature mark of the image.

The signature brand that we have generated is a square matrix of 64 * 64 of real numbers. Figure 6 shows a sample of this matrix which is constructed recursively as follows:

- Creating sub block size of 64 pixels.

- Propagating each block in the ANN and retrieving in the output another block of 4 pixels.
- Reconstitution of the blocks generated by the ANN to form a new image.
- If the image is smaller strictly to 2 * (64 * 64), then our brand signature is the first block of size (64 * 64) retrieved from the image.
- If not, it returns to the first step and we repeated the treatment on the new matrix.

Table1. Example of signature brand matrix

0,3909787	0,5942461	0,1758067	0,570799
0,7838546	0,4328258	0,9424887	0,134017
0,6287956	0,0495958	0,9381059	0,295321
0,8730479	0,6471385	0,7373837	0,33285
0,9164182	0,1429214	0,1170479	0,388731
0,9753208	0,997113	0,2675925	0,2199
0,7721224	0,9261069	0,126968	0,465133
0,3155098	0,3050952	0,8643453	0,404389
0,3844193	0,8711779	0,2627959	0,420844
0,4696244	0,3227052	0,4210994	0,937877
0,7938168	0,2988817	0,6899086	0,672164
0,0246497	0,3149587	0,3639539	0,711749
0,8075495	0,7368181	0,4869043	0,321361

3.2. Direct signature with learning

We used the $ANN\,(64, 16, 4)$ with an unsupervised learning algorithm deducted from the back-propagation of the gradient [8].

Learning principle is based on our concept of image and signatures similarity rate that we have defined in the introduction of this work.

We built the training base by slightly modified images of Lenna (1000 copies). Figure 2 shows an example of these images.

Figure 2. Example images of the learning database

The similarity rate of these two images is very elevated ($T\,(M1, M2) = 0.001$), we evaluate the degree of similarity of the signatures produced by the two images.

The objective of learning is to ensure a higher level of divergence between signatures. Our ultimate goal is to have two completely different signatures for two infinitely similar images so we can prove the strength of our solution against attacks. The learning algorithm is as follow:

1. Random Initialization of weight.
2. Propagation of signal presented on ANN inputs.
3. Propagation of signal in to the hidden layers.
4. Propagation of signal to the output layer.
5. From the second iteration of the algorithm, we calculate the rate of similarity between the current image (the block: SC) and the previous image (the block: ST).
6. If ST> SC then we go directly to the next image (block).
7. If not we go to step 6.

8. Back propagation of error.
9. Updates of knowledge in the weight.
10. Loop to step 2

The graph 3 represents the convergence of learning depending on the iteration number.

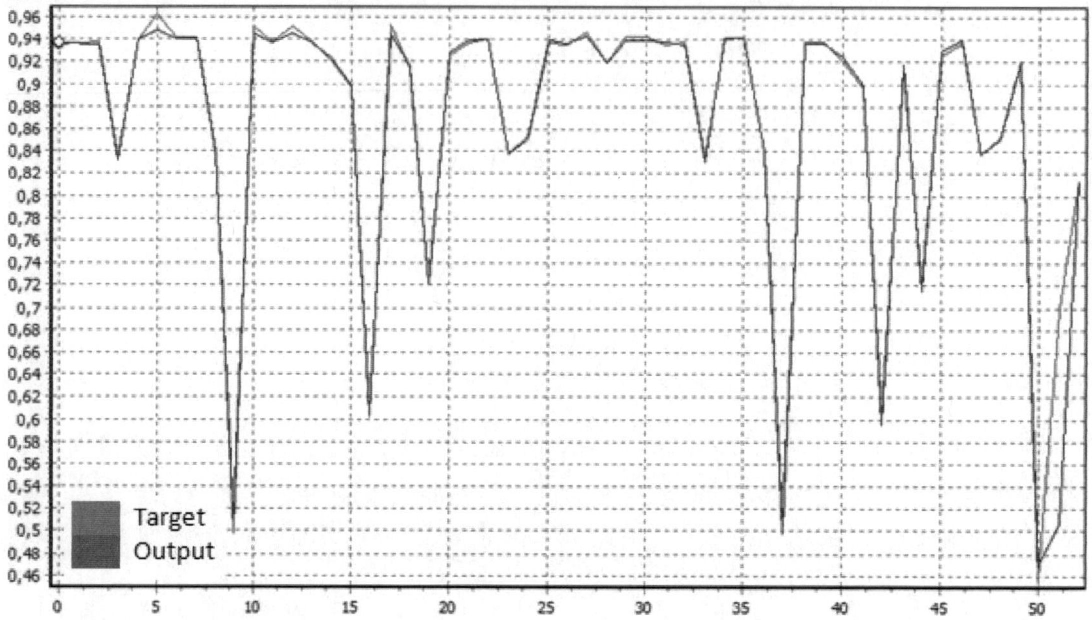

Figure 3. Convergence of learning based on iterations number

4. Results & Discussion

4.1. Direct signature without learning

The desired results are to prove experimentally the relationship between the image and signature similarity rate with the direct signature. Table 2 shows an excerpt from results:

Table 2. Excerpt from results of direct signature without learning

Image and signature similarity rate		Image m1 T(m1, mi)	Image m2 T(m2, mi)	Image m3 T(m3, mi)	Image m4 T(m4, mi)	Image m5 T(m5, mi)	Image m6 T(m6, mi)	Image m7 T(m7, mi)
Image m1	T(m1, mi)=0.001	**1**	0,791633	0,180129	0,073344	0,552294	0,839633	0,333491
Image m2	T(m2, mi)=0.002	0.001	**1**	0,085628	0,502186	0,523391	0,998496	0,988435
Image m3	T(m3, mi)=0.003	0.002	0.001	**1**	0,87703	0,641848	0,208981	0,949327
Image m4	T(m4, mi)=0.004	0.003	0.002	0.001	**1**	0,269716	0,608958	0,687738
Image m5	T(m5, mi)=0.005	0.004	0.003	0.002	0.001	**1**	0,839663	0,421186
Image m6	T(m6, mi)=0.006	0.005	0.004	0.003	0.002	0.001	**1**	0,358951
Image m7	T(m7, mi)=0.007	0.006	0.005	0.004	0.003	0.002	0.001	**1**
	Image similarity rate	S(m1, mi)	S(m2, mi)	S(m3, mi)	S(m4, mi)	S(m5, mi)	S(m6, mi)	S(m7, mi)
	Signature similarity rate	Images and signatures similarity rate with 0<i<1025						

This table is composed of 1,000 Lena images infinitely close. The red boxes represent the similarities rate between two images and the blue boxes to represent the signatures similarities rates for both images.

The graph of Figure 4 shows the relationship between the similarity rate of the images and signatures to our database composed of 1,000 of infinitely close.

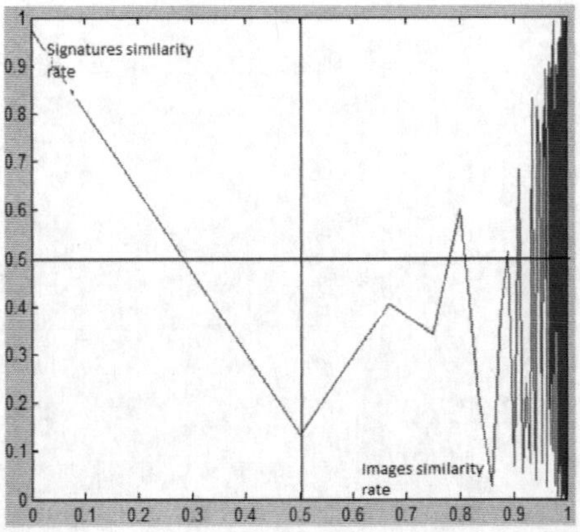

Figure 4. Relationship between the similarity rate of images and signatures (without learning)

Note that this graph is not 100% random, which proves that we have a partial relationship between the similarity rate of the images and signatures. This reflected a major drawback to the approach without learning. Because we can calculate the signature trademark of an image if we have the signature of another relatively close image.

To solve this problem, we used unsupervised learning for multilayer ANN.

4.2. Direct Signature with learning

Table 4 shows a second excerpt of the matrix of calculated results with $ANN\,(64, 16, 4)$ with the unsupervised learning algorithm.

Table 3. Excerpt from results of direct signature with learning

Image and signature similarity rate		Image m1 T(m1, mi)	Image m2 T(m2, mi)	Image m3 T(m3, mi)	Image m4 T(m4, mi)	Image m5 T(m5, mi)	Image m6 T(m6, mi)	Image m7 T(m7, mi)
Image m1	T(m1, mi)=0.001	**1**	0,791633	0,180129	0,073344	0,552294	0,839633	0,333491
Image m2	T(m2, mi)=0.002	0.001	**1**	0,085628	0,502186	0,523391	0,998496	0,988435
Image m3	T(m3, mi)=0.003	0.002	0.001	**1**	0,87703	0,641848	0,208981	0,949327
Image m4	T(m4, mi)=0.004	0.003	0.002	0.001	**1**	0,269716	0,608958	0,687738
Image m5	T(m5, mi)=0.005	0.004	0.003	0.002	0.001	**1**	0,839663	0,421186
Image m6	T(m6, mi)=0.006	0.005	0.004	0.003	0.002	0.001	**1**	0,358951
Image m7	T(m7, mi)=0.007	0.006	0.005	0.004	0.003	0.002	0.001	**1**
	Image similarity rate	S(m1, mi)	S(m2, mi)	S(m3, mi)	S(m4, mi)	S(m5, mi)	S(m6, mi)	S(m7, mi)
	Signature similarity rate	Images and signatures similarity rate with 0<i<1025						

Figure 5 shows the relationship between the similarity rate of the images and signatures for images we studied.

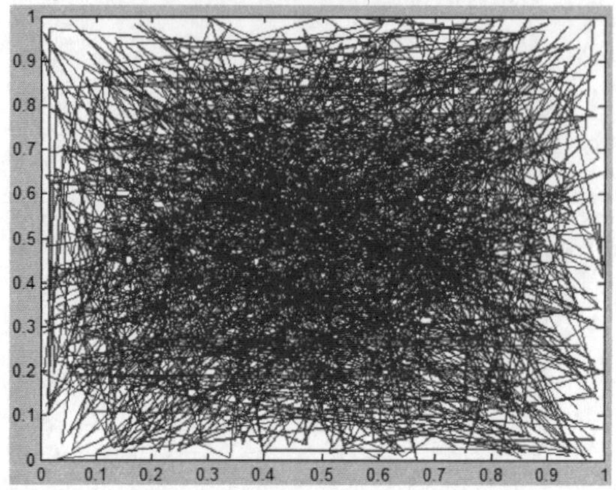

Figure 5. Relationship between the similarity rate of images and signatures (with learning)

The graph in figure 5 is completely random. We can conclude that this approach cannot provide the same signature for both images regardless of their degree of similarity.

5. Conclusion

We have presented in this work, contributions that use ANN in different ways to protect the image integrity. We found that with unsupervised learning, we cannot have two identical signatures for two images, whatever their degree of resemblance. This new approach is very important because for two almost identical images, one can check which one is original and which one is falsified.

Our approach has several advantages:

- It offers both a reliable signature and verification method.
- It is independent of any signature or verification key.
- It is independent of the form of the signed image.
- The size of the signature is the same and does not depend on the image size.
- It is resistant to various attacks.

However, the main inconvenience of this approach is the difficulty of the use of ANN especially the relatively large learning time.

References

[1] F. Cayre, C. Fontaine, & T. Furon, « Watermarking security: theory and practice », IEEE Trans. Signal Process., vol. 53, no 10, p. 3976- 3987, 2005.

[2] H. Baltzakis & N. Papamarkos, « A new signature verification technique based on a two-stage neural network classifier », Eng. Appl. Artif. Intell. vol. 14, no 1, p. 95- 103, 2001.

[3] J.-P. Drouhard, R. Sabourin, & M. Godbout, « A neural network approach to off-line signature verification using directional PDF », Pattern Recognit., vol. 29, no 3, p. 415- 424, 1996.

[4] S. Y. Ooi, A. B. J. Teoh, Y. H. Pang, et B. Y. Hiew, « Image-based handwritten signature verification using hybrid methods of discrete radon transform, principal component analysis and probabilistic neural network », Appl. Soft Comput., vol. 40, p. 274- 282, 2016.

[5] O. C. Abikoye, M. A. Mabayoje, & R. Ajibade, « Offline signature recognition & verification using neural network », Int. J. Comput. Appl., vol. 35, no 2, p. 44- 51, 2011.

[6] M. V. Pandey & S. Shantaiya, « Signature verification using morphological features based on artificial neural network », Int. J. Adv. Res. Comput. Sci. Softw. Eng., vol. 2, no 7, 2012.

[7] P. Shikha & S. Shailja, « Neural network based offline signature recognition and verification system », Res. J. Eng. Sci. ISSN, vol. 2278, p. 9472, 2013.

[8] K. Ben & P. Van der Smagt, « An introduction to neural networks », Univ. Amst., 1996.

IOP Conf. Series: Materials Science and Engineering **353** (2018) 012017 doi:10.1088/1757-899X/353/1/012017

Incidents Prediction in Road Junctions Using Artificial Neural Networks

Tarik HAJJI, Aicha AlAMI HASSANI and Mohammed OUAZZANI JAMIL

Laboratoire Systèmes et Environnements Durables (SED), Faculté des Sciences de l'Ingénieur (FSI), Université Privée de Fès (UPF), Fez, Morocco

hajji@upf.ac.ma

Abstract. The implementation of an incident detection system (IDS) is an indispensable operation in the analysis of the road traffics. However the IDS may, in no case, represent an alternative to the classical monitoring system controlled by the human eye. The aim of this work is to increase detection and prediction probability of incidents in camera-monitored areas. Knowing that, these areas are monitored by multiple cameras and few supervisors. Our solution is to use Artificial Neural Networks (ANN) to analyze moving objects trajectories on captured images. We first propose a modelling of the trajectories and their characteristics, after we develop a learning database for valid and invalid trajectories, and then we carry out a comparative study to find the artificial neural network architecture that maximizes the rate of valid and invalid trajectories recognition.

Keywords. Incidents prediction, Images and videos processing, ANN, Artificial intelligence.

1. Introduction

The installation of surveillance cameras is an essential operation for the security, the detection of fraud, thefts, burglaries, assaults, vandalism. However, this operation is necessary but insufficient to meet the requirements of the monitoring. Consequently, the installation of an additional software layer like an IDS is highly recommended to increase the monitoring system performances.

The French Tunnel Research Engineering published in 2015 a report entitled «automatic incident detection using image analysis in the tunnel» [1]. The report contains general information and history of IDS, different types of incidents to detect the description of the operating principle such as the arrangement of cams and disruption management, the operating mode in the case of detection of an incident by the IDS valuation method.

1.1. Artificial neural networks

ANN are problem-solving tools inspired by the biological neural networks that constitute human brains, and it's a network of many very simple processors ("units"), each possibly having a local memory. It is based on a collection of connected artificial neurons organized in layers. Most ANN have three layers: the input layer, a hidden layer, and the output layer.

IOP Conf. Series: Materials Science and Engineering **353** (2018) 012017 doi:10.1088/1757-899X/353/1/012017

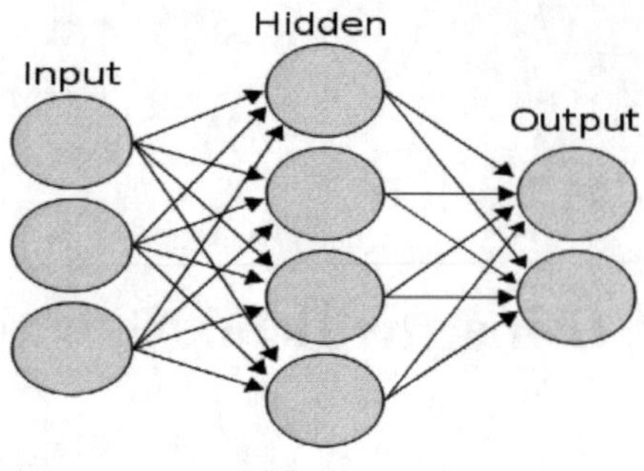

Figure 1. ANN[2]

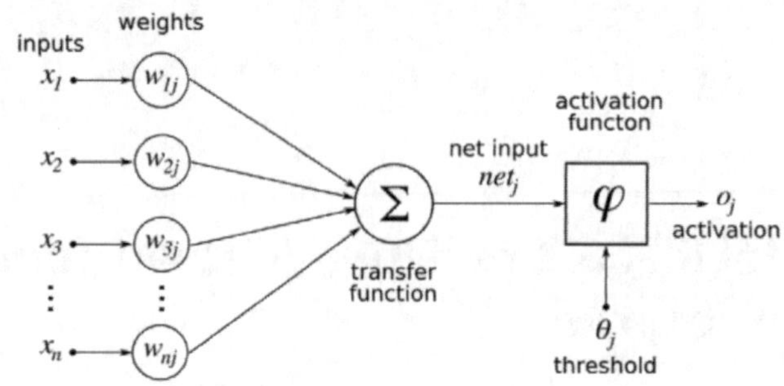

Figure 2. Artificial neural[2]

(Fig. 1) shows the structure of an ANN containing one hidden layer and one output layer with 4 neurons and 2 neurons, respectively, and three inputs. An artificial neural is a function able to realize a non-linear sum of its inputs (Fig. 2.). Most ANN have some sort of "training" rule whereby the weights of connections are adjusted on the basis of presented patterns. The Back-Propagation algorithm [2] which is one of the most popular learning algorithms is a method used in ANN to calculate the error contribution of each neuron after a batch of data is processed.

2. State-of-the-art

The aim of this part is to make an overview of the various methods of abnormal event detections in videos. The analysis of moving objects behavior in videos attracts a great number of researchers and therefore several approaches have been proposed. Most of them deem movement as low-level information. They then define intermediate level descriptors to model the studied behavior.

2.1. Foreground obtained using a background extraction process:

The simplest way to separate foreground from background is to acquire a background image that does not contain any moving object. A subtraction between each new image and initial background image is then performed with respect to a threshold. However a single image of background is often not available because it changes continuously due to various events. The background modeling is used in a variety of applications such as video surveillance [3, 4] or multimedia [5].

Methods can be classified into the following categories: background basic modeling [6], background statistical modeling [7], background fuzzy modeling and background estimation [8]. They can be also classified according to prediction [9], recursion [3], adaptation [10], or modality [11].

The Gaussians are then heuristically marked as follows: the darkest component is marked as a shadow and the other two components are differentiated according to the variance (the greatest variance is labelled as a vehicle and the other as the road). The foreground classification is made by associating each pixel with the corresponding Gaussian. An incremental algorithm is used for background maintenance to perform real-time processing. However, this process suffers from a lack of adaptation to the changes that appear in the scene through time. Stauffer and Crimson [12] have generalized this idea by modeling the color intensity of each pixel over the previous time interval *{X1, X2, Xt}* by a mixture of Gaussians.

2.2. Points of interest:

Points of interest are pixels of the image that are likely to be tracked effectively in the following frames. The Harris algorithm for interest point's extraction [13] is known for its invariance.

3. Methodology

Our approach consists in starting with the modeling of a moving object (MO) by a motion vector composed of a collection of characteristics representing its behavior in the video. Vehicles or pedestrians or other objects captured by the cameras in the monitored area are referred to as movable objects. A Motion Vector (MV) of a MO also designates a mathematical object capable of representing the state of a moving object at a given instant. Then we build a Knowledge Database (KDB) formed by MV for different MO and the score of each MV. The KDB is used as a data source for the system-learning algorithm. Afterwards, a comparative study is made to select the most suitable ANN architecture for this subject in terms of recognition rate and complexity. Finally, the method is evaluated and the results are compared with other SDIs.

3.1. Design of the motion vector

In order to characterize a MO at a given instant, a set of characteristics of the object's movement is extracted. In this study, we consider the following characteristics: the mask, the speed, the acceleration, the type of the trajectory, the direction, and the volume. These characteristics represent standardized mathematical quantities:

- The mask indicates the number of the area occupied by the moving object on the capture
- The type of the trajectory indicates the moving object type of movement: rectilinear uniform, rectilinear, sinusoidal, diagonal, circular ...
- The direction indicates the movement direction, it is a number between 0 and 386.
- Speed and acceleration are two classical physical quantities
- The volume indicates the area occupied by the moving object on the capture.

Thus the motion vector for an object is defined as follows *Vm = [mask, velocity, acceleration, trajectory, direction, volume]*. The figure below shows an example of a motion vector for a moving object:

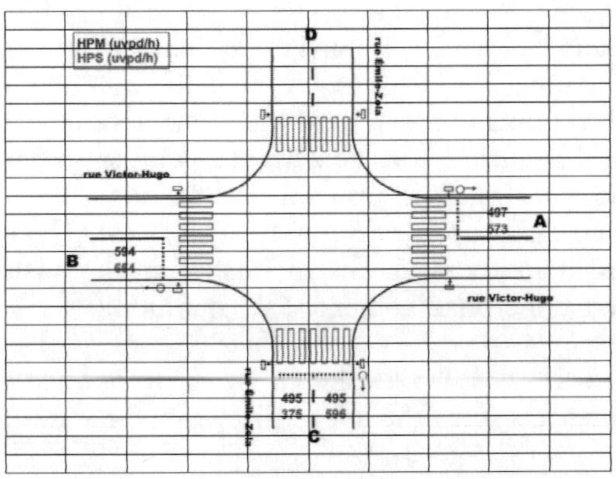

Figure 3. Moving objects: A, B, C and D

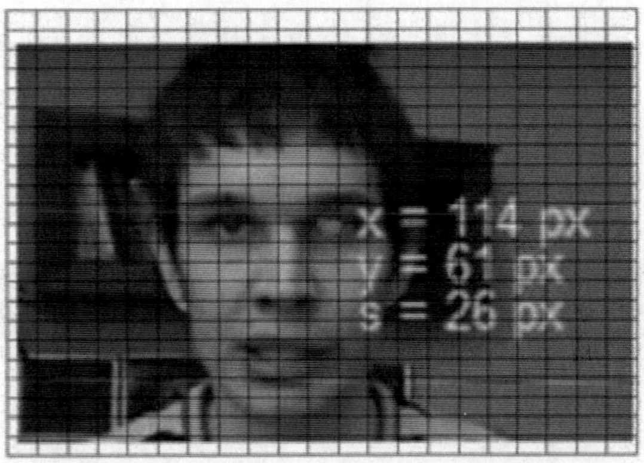

Figure 4. The mask deduction of the MO

The (Fig. 3) is composed of 4 moving objects A, B, C and D. The motion vector for each of them is respectively: *Vm(A)=[4,60, 0.5, 2, 180, 10] ; Vm(B)=[4,60, 0.5, 2, 180, 10] ; Vm(C)=[4,60, 0.5, 2, 180, 10] ; Vm(D)=[4,60, 0.5, 2, 180, 10] ;*

3.1.1. Mask of a moving object. The mask of a moving object is an identifier with a specific area on the capture. The zones are numbered from 1 to n, where n is the number of mask, in figure n equals to 200. In the case where the mobile object is distributed over more than one mask, then the most occupied mask is considered. We use *OpenTLD-master* (TLD) [14] which is an algorithm for tracking unknown objects in unconstrained video streams. The object of interest is defined by a bounding box in a single frame. TLD follows simultaneously the object, learns its appearance and detects it whenever it appears in the video. The algorithm is implemented by **Z. KALAL** in form of a Matlab library, and then we extract the coordinates of the MO on the scene captured by the camera and infer from it the mask. The (Fig. 4) explains the operating mechanism for the mask deduction of the MO (Red dot), which is the eye of the **Z. KALAL**.

3.1.2. Moving object trajectory. The trajectory of a moving object designates the motion type of an MO. We have considered a continuous notation of the moving object trajectories: 1 for uniform rectilinear, 2 for rectilinear, 3 for sinusoidal, 4 for diagonal and 5 for circular. We used the notion of polynomial interpolation [15] for the classification of the MO trajectories in a mask.

The principle of the technique is to detect the trajectory curve according to the occupations of the moving object in a mask. The occupations are represented by the moving object coordinates in relation to a given reference. We compute these coordinates by *OpenTLD-master*.

For the recognition of the trajectory type represented by the polynomial, we used an ANN with supervised learning on a set of points of the predefined trajectories. The trajectories studied in this work, each one of them is represented by a Lagrange polynomial. These polynomials allow us to build a learning database on a set of points. For each polynomial we compute the images of a collection of n points $X_i = P_j [X_i]$ (P_j represents the polynomial of the trajectory j and X_i represents the point i) and a learning database is obtained for the ANN. The (Tab. 1) shows an extract of this database.

Table 1. Learning database for the trajectories classification

X_i	$P_1[x_i]$	$P_2[x_i]$	$P_3[x_i]$	$P_4[x_i]$	$P_5[x_i]$
0,90576579	0,52169653	0,10916584	0,33935151	0,3173368	0,37169586
0,61152489	0,45771693	0,46002726	0,55990782	0,92800427	0,26958747
0,72269234	0,93201435	0,83723601	0,72172925	0,69169074	0,24253564

We use an ANN with an input layer of 6 artificial neurons, a hidden layer of 3 artificial neurons and an output layer of 5 artificial neurons and the backpropagation algorithm for learning.

3.1.3. Moving object speed. It is a mathematical quantity, which designates the number of pixels traversed by a moving object in a time interval (pixels / time). We have experimentally fixed the time interval required to calculate the speed of a moving object on a mask. We exploited the coordinates of the mobile object returned instantaneously by *OpenTLD-master* to calculate the number of pixels travelled by the moving object during a fixed time interval. The principle of the algorithm for calculating the moving object speed is simple; we saved the couple (position, time) in a list. Then we considered that the speed of the object during a fixed duration is the distance between the position recorded at the initial moment and the position recorded at the final moment of the duration.

3.1.4. Moving object acceleration. This parameter represents the evolution of the mobile object movement in a mask. It is a mathematical quantity that measures the speed difference between two instants.

3.1.5. Moving object direction. We designate by this mathematical magnitude the angle between the motion direction of the moving object and the horizontal axis used as a reference on a segment on a mask.

3.1.6. Moving object volume. Several solutions are available and able to calculate the actual dimensions of an object on an image captured as *UPHOTOMEASURE* [16]. In this work we considered only the virtual volume of the moving object and not the actual volume occupied in the real space.

3.2. Design of the knowledge base

We have experimentally built this knowledge base (Tab. 2) in our laboratory. We created a prototype of a roundabout, used a camera to monitor and simulate moving objects by objects that we have prepared in our laboratory. Then we programmed valid and invalid configurations and finally computed the motion vectors for each moving object.

Table 2. Base of knowledge

MO	Mask	Trajectory	Speed	Acceleration	Direction	Surface	Score

IOP Conf. Series: Materials Science and Engineering **353** (2018) 012017 doi:10.1088/1757-899X/353/1/012017

1	625	9	145	5	195	97	3
2	47	6	129	-4	223	24	4
3	58	3	58	-3	136	4	1
4	417	7	35	-8	0	73	4
5	75	6	116	6	67	78	6

3.3. Implementation:
We have developed a JAVA application that generates an ANN with the postulated architecture based on a simple description in the form of an xml meta-model. The application also has the ability to start learning of the ANN by the back-propagation of the gradient. The learning data are loaded via input.txt and output.txt files and the learning results are generated in the results.txt file.

4. Results
The (Tab. 3) shows the results of a comparative study to choose the structure of the most adopted ANN for this subject in terms of convergence rate.

Table 3. Comparative study results

ANN	Quadratic Error	Number of iterations	Recognition Rate Learning Database (%)	Recognition Rate Test Database (%)
6 :3 :1	**0.01**	10^6	100	97
		10^9	100	99
		$\mathbf{10^{10}}$	**100**	**99,8**
	0.0001	10^6	100	98
		10^9	100	99,6
		10^{10}	100	100
6 :2 :1	0.01	10^6	100	97
		10^9	100	99
		10^{10}	100	99,8
	0.0001	10^6	100	98
		10^9	100	99,6
		10^{10}	100	100
6 :3 :2 :1	0.01	10^6	100	97
		10^9	100	99
		10^{10}	100	99,8
	0.0001	10^6	100	98
		10^9	100	99,6
		10^{10}	100	100

Based on these results, we can conclude that with the ANN characterized by: an input layer composed of 6 artificial neurons, a hidden layer composed of 3 artificial neurons, an output layer composed of a single artificial neuron, a quadratic error average of the order of 10^{-4} and with an iteration number equal to 10^{10} we have a perfect recognition rate equal to 100% tested on the learning database.

5. Conclusion
In this work, we deal with the subject of automatic detection of undesirable events in surveillance cameras in the roundabouts of cities in order to propose intelligent, rapid and effective intervention. We proposed a solution based on artificial neural networks with a supervised learning for the

classification of detected undesirable events. These events are detected via a vector value calculated in real time on the videos of the surveillance cameras. The experimental results are very encouraging and our perspective is to evaluate this approach on a real case.

References

[1] Centre d'Etudes des Tunnels, détection automatique d'incident par analyse d'images en tunnel, France, 2015, http://www.cetu.developpement-durable.gouv.fr/IMG/pdf/CETU_DocInfo_DAI_2015.pdf
[2] Syed Danish Ali and Rahul Ahuja, The Evolution and Core Concepts of Deep Learning & Neural Networks, AUGUST 3, 2016, https://www.analyticsvidhya.com/blog/2016/08/evolution-core-concepts-deep-learning-neural-networks/.
[3] Cheung and . Kamath, Robust Background Subtraction with Foreground Validation for Urban Traffic Video, Special Issue on Advances in Intelligent Vision Systems : Methods and Applications, pp.2330-2340, 2005.DOI10.1155/ASP.2005.2330.
[4] Jiyun Zhai, Xin Zhou, Congqing Wang, "A moving target detection algorithm based on combination of GMM and LBP texture pattern", Guidance Navigation and Control Conference (CGNCC) 2016 IEEE Chinese, pp. 1057-1060, 2016.
[5] A.PANDE, A. VERMA et A. MITTAL : Network aware optimal ressource allocation for e-learning videos. In 6th International Conference on mobile Learning, Melbourne- Australia, 2007.
[6] Lee, B.; Hedley, M. Background estimation for video surveillance. In: Image & Vision Computing New Zealand (IVCNZ '02); 2002; Auckland, NZ. 2002. 315-320.
[7] Xuannan Ye, Wanggen Wan, "Fast background modeling using GMM on GPU", Audio Language and Image Processing (ICALIP) 2014 International Conference on, pp. 937-941, 2014.
[8] S. Messelodi, C. Modena, N. Segata, and M. Zanin, A Kalman Filter Based Background Updating Algorithm Robust to Sharp Illumination Changes, 13th International Conference on Image Analysis and Processing (ICIAP), pp.163-170, 2005. DOI10.1007/11553595_20
[9] Wang and . Suter, A Novel Robust Statistical Method for Background Initialization and Visual Surveillance, Asian Conference on Computer Vision (ACCV), pp.328-337, 2006. DOI10.1007/11612032_34
[10] F. Por03 and . Porikli, Human body tracking by adaptive background models and meanshiftanalysis, IEEE International Workshop on Performance Evaluation of Tracking and Surveillance, 2003.
[11] Z. Tao and N. Ram, Bayesian human segmentation in crowded situations, IEEE Computer Society Conference on Computer Vision and Pattern Recognition (CVPR), p.459, 2003.
[12] Stauffer and . Grimson, Adaptative background mixture models for real-time tracking, IEEE Conference on Computer Vision and Pattern Recognition (CVPR), pp.246-252, 1999.
[13] Mrinal Kanti Bhowmika, Sandip Shilb, Priya Sahac: Feature Points Extraction of Thermal Face using Harris Interest Point Detection: International Conference on Computational Intelligence: Modeling Techniques and Applications(CIMTA) 2013.
[14] Z.Kalal, k.Mikolajczyk and J.Matas: Tracking-learning-detection, Journal Pattern Analysis and Machine Intelligence, 2012.
[15] Endre Süli and David Mayers (2003). An Introduction to Numerical Analysis, Chapter 6. Cambridge University Press. ISBN 0-521-00794-1.
[16] Digicontractor Corporation:http://www.uphotomeasure.com/UPM_a2/index.htm, 2016.

IOP Conf. Series: Materials Science and Engineering **353** (2018) 012018 doi:10.1088/1757-899X/353/1/012018

Development of a cellulose-based insulating composite material for green buildings: Case of treated organic waste (paper, cardboard, hash)

Ahmed OUARGUI[1], Naoual BELOUAGGADIA[1], Abdeslam ELBOUARI[2] and Mohammed EZZINE[2]

[1] Laboratory of Engineering and Materials (LIMAT), Ben M'Sik Faculty of Sciences, Hassan II University of Casablanca, B.P. 7955 Sidi Otmane, Casablanca, Morocco
[2] Laboratory of Physico-Chemistry of Applied Materials (LPCMA), Ben M'Sik Faculty of Sciences, Hassan II University of Casablanca, B.P. 7955 Sidi Otmane, Casablanca, Morocco.

Email: ahmedouargui@gmail.com

Abstract. Buildings are responsible for 36% of the final energy consumption in Morocco [1-2], and a reduction of this energy consumption of buildings is a priority for the kingdom in order to reach its energy saving goals. One of the most effective actions to reduce energy consumption is the selection and development of innovative and efficient building materials [3]. In this work, we present an experimental study of the effect of adding treated organic waste (paper, cardboard, hash) on mechanical and thermal properties of cement and clay bricks. Thermal conductivity, specific heat and mechanical resistance were investigated in terms of content and size additives. Soaking time and drying temperature were also taken into account. The results reveal that thermal conductivity decreases as well in the case of the paper-cement mixture as that of the paper-clay and seems to stabilize around 40%. In the case of the composite paper-cement, it is found that, for an additives quantity exceeding 15%, the compressive strength exceeds the standard for the hollow non-load bearing masonry. However, the case of paper-clay mixture seems to give more interesting results, related to the compressive strength, for a mass composition of 15% in paper. Given the positive results achieved, it seems possible to use these composites for the construction of walls, ceilings and roofs of housing while minimizing the energy consumption of the building.

1. Introduction

The economy of the energy is increasingly a major challenge in the world because of economic and ecological concerns. Globally, the building sector alone accounts for 32% of final energy consumption and accounts for around a third of CO_2 emissions [3-4].

In Morocco, the energy consumption generated by buildings represents around 36% of total final energy consumption [5-6]. Reducing the energy consumption of buildings has therefore become a priority for the kingdom to achieve its greenhouse gas reduction targets. Among the priority actions, emphasis focused on selection and development of innovative building materials with interesting thermal and mechanical properties Similarly, the conditions of thermal comfort in a room also depends on the activity in the neighboring premises. In the case of industrial activities, this generates thermal conditions that are difficult to mitigate by the construction products normally used in the

IOP Conf. Series: Materials Science and Engineering **353** (2018) 012018 doi:10.1088/1757-899X/353/1/012018

separation walls. This leads to large variations in the temperature of the surrounding premises and requires significant energy to maintain them at their thermal comfort level. For example, the Moroccan Construction Thermal Regulation (RTCM) states that, depending on the location and climate, the walls should be made of material with a heat transfer coefficient of between 0.55 and 1.20 W / m2 K [6-7]. In this work, we have chosen as raw material a by-product of paper to be integrated with cement or clay in order to reinforce their quality of thermal insulation. The mechanical and thermal properties of the composite blocks were studied according to the amount of additives, the time of soaking in water and the drying temperature.

2. Material and experimental method

2.1. Materials and implementation

Our work consists of developing two types of paper-based composite materials. These two composites will be subjected thereafter to thermal and mechanical tests.

For the first prototype called "paper cement», the main constituent elements are paper waste mixed with cement, sand and water. Each component itself was subjected to initial preparation, starting with mixing with water followed by purification before mixing. The mixing process allows the transformation of waste paper into granulates. The latter will be wet before being dried at room temperature. For the second prototype, it consists of the following constituents: clay, paper waste, sand and water. The preparation protocol is the same as that of the "paper cement".

2.2. Geometry and conditioning of the samples

The "cement-paper" and "clay-paper" plates were prepared by molding in the form of a square of 20cm of side with a thickness of 2 cm. The following table gathers the physical and geometrical characteristics of the various constituents of the composites.

Table 1.

Component	Mass volume g /cm3	Range (mm)	Thickness Plate (cm)	L=l(Cm)
Paper	0,77	1.5		
Clay	1,7	1.5		
Sand	1,6	1	2	20
Cement	1.15	--		
Water	1	--		

Density of the constituents of the composites.

The plates thus prepared were made in order to test the thermal conductivity. For mechanical bending tests, specimens were molded and prepared; these have the following characteristics: an area of (4*4 cm^2) with a length of 16 cm. Both types of specimens were then dried for 2 hours in an oven at a temperature of T = 105 ° C. Finally, the whole was treated in an oven at a temperature of 300 ° during. Both types of specimens were then dried for 2 hours in an oven at a temperature of T = 105 ° C. Finally, the whole was treated in an oven at a temperature of 300 ° during.

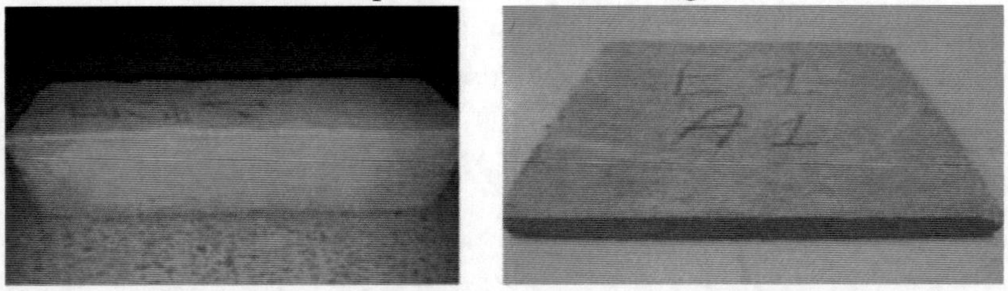

Figure 1 Specimens made by addition of paper treated with cement and clay.

The methodology followed in our study consists in following the thermal conductivity and the mechanical rigidity of the samples prepared according to the ratio of additives. For this, the tests carried out consisted of samples using the following proportions of paper (0%, 5%, 7%, 10%, and 15%), with a fixed percentage (10%) of sand and a varying one (25-30%) of water.

Table 2. Composition of the plates prepared according to the mass composition of paper.

		Paper-Cement			
Plates	Components	%	ρ(g/cm3)	V (cm3)	m(g)
P1	Paper	0	0.77	0	0
	Cement	100	1.15	800	920
P2	Paper	5	0.77	40	31
	Cement	55	1.15	440	506
P3	Paper	7	0.77	56	44
	Cement	53	1.15	424	490
P4	Paper	10	0.77	80	62
	Cement	50	1.15	400	463
P5	Paper	15	0.77	120	92.4
	Cement	45	1.15	360	414

Table 3. Composition of specimens prepared according to paper mass composition for flexural tests.

		Paper-Cement			
Specimens	Components	%	ρ(g/cm3)	V (cm3)	m(g)
E1	Paper	0	0.77	0	0
	Cement	100	1.15	256	295
E2	Paper	5	0.77	13	10
	Cement	55	1.15	138	159
E3	Paper	7	0.77	18	14
	Cement	53	1.15	136	157
E4	Paper	10	0.77	26	21
	Cement	50	1.15	129	149
E5	Paper	15	0.77	39	31
	Cement	45	1.15	116	134

The thermal conductivity measurements were performed for the "cement-paper" and "clay-paper" plates after an adequate treatment of their surfaces. All the tests were carried out with a λ-Meter EP 500

IOP Conf. Series: Materials Science and Engineering **353** (2018) 012018 doi:10.1088/1757-899X/353/1/012018

apparatus according to the method of the guarded hot plate. Measurements were carried out at a temperature of 25 ° C, with a temperature difference of 10 K between the plates. The steady state is considered to be reached when the conductivity varies by less than 1% during a time interval of 60 minutes.

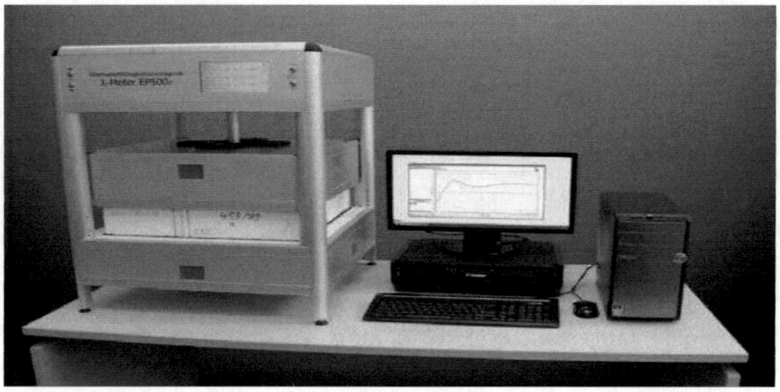

Figure 2 The hot plate machine keeps λ-METER EP500e [8].

For rectangular specimens, a flexural test was used to evaluate the maximum stress at breat, the latter was determined by three-point bending on bars of length l = 20 cm and thickness (e=2cm). Maximum stress was achieved using a KERNEL SISTEML brand machine (0 + 400 kg), (see figure 3). The test piece is held on two simple supports, distant of L = 16 cm. A charge Fr is applied at a point equidistant from the two supports.

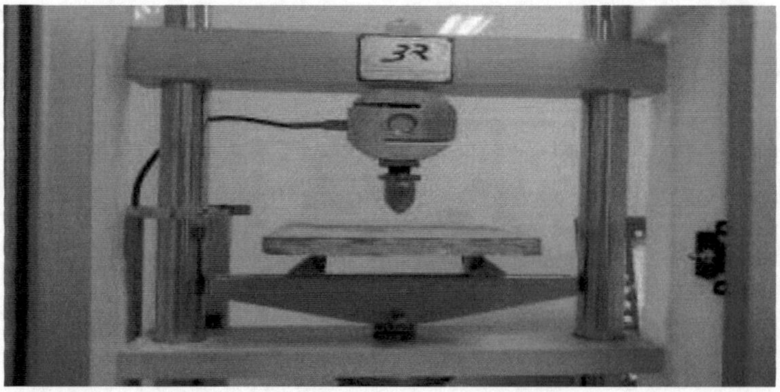

Figure 3 Machine brand KERNEL SISTEML (0 + 400kg) [9].

3. Results and discussion

The results of thermal and mechanical measurements made on the "cement-paper" and "clay-paper" plate are summarized in table 4:

Table 4. Evolution of the thermal conductivity according to the mass composition of paper.

		Paper	P1	P2	P3	P4	P5
Paper-cement	λ(W/m.K)	0.068	0.218	0.195	0.183	0.140	0.107
Paper -clay	λ(W/m.K)	0.068	0.244	0.203	0..152	0.134	0.101

We have noted that the thermal conductivity of both "paper cement" and "paper-clay" specimens increases steadily as the percentage of paper increases in the matrix. It reaches about 40% when the weight ratio paper / cement is 15%. Similarly, the mechanical tests were carried out on a Hounsfield H50KS press, the bending and compression tests were carried out at the respective displacement speeds of 0.3 mm / min and 0.5 mm / min. Two samples were tested for every deadline. The flexural performance of the composites reinforced by the treated paper was determined (see Table 5). It can also be noted that the breaking force, FR and the breaking stress σr, decrease regularly as the percentage of paper increases in the cement. On the other hand, these two properties are improved in the case of clay; which suggests that in the latter case, the paper incorporation leads to the improvement of the mechanical behavior of the composite, increasing its ductility compared to the fragile behavior of the matrix alone. On the other hand, these two properties are improved in the case of clay. This suggests that in the latter case, the paper incorporation leads to the improvement of the mechanical behavior of the composite, increasing its ductility compared to the fragile behavior of the matrix alone.

Table 5. Evolution of the mechanical properties of the specimens (FR: breaking force, σr: breaking stress, S: surface of the specimens)

Specimens		Paper	E1	E2	E3	E4	E5
Paper - Cement	FR (KN)	--	7.84	6.51	5.01	4.40	4.14
	S (cm2)	--	56	56	56	56	56
	σr (Mpa)	FR/S	1.4	1.16	0.89	0.79	0.73
Paper - Clay	FR (KN)	--	4.28	3.80	4.01	4.65	5.01
	S (cm2)	--	56	56	56	56	56
	σr (Mpa	FR/S	0,76	0,68	0,71	0,83	0,89

4. Conclusion

The main objective of this work is to evaluate the influence of the different percentages of paper on two clay components and cement in order to undertake a benchmarking study of these two composites,

IOP Conf. Series: Materials Science and Engineering **353** (2018) 012018 doi:10.1088/1757-899X/353/1/012018

and to determine the best insulator among its two materials, based on the parameters bending strength and thermal conductivity. The results obtained show that the "cement-paper" matrix composites are less resistant to fracture and the conductivity. However, the composite "clay-paper" paper has a higher and more reliable conductivity for use in thermal insulation. The results obtained show that the "cement-paper" matrix composites are less resistant to fracture and the conductivity against the composite "clay-paper" paper has a higher and more reliable conductivity for use in thermal insulation.

Abaqus software modeling counts among our future projects to compare the experimental results and numerical ones. In addition, we intend to make flow variation tests on the samples in a machine as a building containing a source of heat inside in order to study conduction transfer on walls.

References

[1] R. Snellings, Ö. Cizer, L. Horckmans, P.T. Durdzin ´ ski, P. Dierckx, P. Nielsen, K.Van Balen, L. Vandewalle, (2016) 35–39, Properties and pozzolanic reactivity of flash calcined dredging sediments, Appl. Clay Sci. 129, http://dx.doi.org/10.1016/j.clay.2016.04.019.

[2] Mahfoud Benzerzour , Mouhamadou Amar and Nor-Edine Abriak (2017), New experimental approach of the reuse of dredged sediments in a cement matrix by physical and heat treatment, Construction and Building Materials 140,432–444

[3] P. Wouter, 2004, Energy Performance of Building: Assessment of Innovative Technologies, ENPER- TEBUC, Final Report.

[4] EN 832:1998—Thermal Performance of Buildings—Calculation of Energy Use for Heating—Residential Buildings, CEN, Brussels, Belgium,1998.More references.

[5] 2004,American society for testing and materials Standard, Test Method for Steady-State Heat Flux Measurements and Thermal Transmission Properties by Means of the Guarded-Hat-Plate Apparatus.

[6] P. IEA/UNDP Modernizing building energy codes IEA 2013.

[7] ADEREE 2011, The technical elements of the project of the thermal regulation of the building in Morocco

[8] ADEREE, Thermal Regulation of Construction in Morocco (RTCM).

[9] E. Gratia, 2003, A. De Herde Design of low energy office buildings Energy Build 35-(5), pp. 473-491.

[10] The hot plate machine keeps λ-METER EP500e (20/02/2018):[Internet]. Available https://www. the hot plate machine keeps λ-METER EP500e -Hp-500e-atf-p286027.html

[11] Machine brand KERNEL SISTEML (15/02/2018) ,Presse Universelle : RP 25 ATF : [Internet]. Available https://www.usinenouvelle.com/expo/presse-universelle-rp-25-atf-p286027.html

IOP Conf. Series: Materials Science and Engineering **353** (2018) 012019 doi:10.1088/1757-899X/353/1/012019

SMART Grid Evaluation Using Fuzzy Numbers and TOPSIS

Mohammed EL ALAOUI [1,2]

[1] ENSAM Meknes, Moulay Ismail University, Morocco
[2] Laboratoire Systèmes et Environnements Durables (LSED), Faculté des Sciences de l'Ingénieur (FSI), Université Privée de Fès (UPF), Fez, Morocco

mohamedelalaoui208@gmail.com

Abstract. In recent advent of smart grids, the end-users aims to satisfy simultaneously low electricity bills, with a reasonable level of comfort. While cost evaluation appears to be an easy task, capturing human preferences seems to be more challenging. Here we propose the use of fuzzy logic and a modified version of the TOPSIS method, to quantify end-users' preferences in a smart grid. While classical smart grid focus only on the technological side, it is proven that smart grid effectiveness is hugely linked to end-users' behaviours. The main objective here, is to involve smart grid users in order to get maximum satisfaction, preserving classical smart grid objectives.

1. Introduction

It is evident that actual power grid are under effective, whether in economical or environmental terms. The transition to smart grids seems to be a necessity more than a simple choice. This transition can be facilitated, given the number of intelligent measurement and control tools in the market. However, the main challenges are caused by the human activities. Hence, one of the main objectives, is to make end-users collaborate, especially in the period of peak demand, to consume electricity more efficiently [1].
Similarly to a classical supply chain, smart grids must take into consideration the end users' preferences. Several authors insisted in the role of users in the smart grid [2]–[5]. We can program the dishwasher, the cloth drier and other machines in order to function in off-peak periods. However, power consumption still depend mainly on end-users' preferences and life style.
The main question is: "how to model users' preferences?" Fuzzy set theory was first proposed by L. Zadeh [6] in 1965 in order to meet with human reasoning. Then, fuzzy theory was adopted in the decision making framework [7]. Several authors adopted the fuzzy decision making methodology for smart grids [8]–[11]. And employed Technique for Order Preference by Similarity to Ideal Solution (TOPSIS) to rank preferences.
Section 2 gives a brief overview about smart grids .Section 3 presents the needed background of fuzzy logic. Section 4 presents TOPSIS. Section 5 gives an illustrative example. Section 6 concludes this work.

2. Smart grids

It is well known that the main problematic in smart grids is due to the peak load. Some countries, in the hope to influence their citizens' electricity consumption, imposes a varying price, where the maximum price corresponds to the on-peak. So, to reduce the electricity bills, the users are invited to

IOP Conf. Series: Materials Science and Engineering **353** (2018) 012019 doi:10.1088/1757-899X/353/1/012019

shift their consumption from on-peak, to off-peak periods. Unfortunately, this move can result on end user discomfort. Authors in [12] represented the dissatisfaction function depending on some variables such as the difference between the desired temperature and the obtained temperature. However, the dissatisfaction, cannot be reduced, to linearly variables such as temperature or any other measurable variables. Hence, all crisp models will fail to capture human judgments.

3. Fuzzy logic

To deal quantitatively with the ambiguity of human judgement, L. Zadeh introduced fuzzy sets [6]. Let X be a universal set and F a fuzzy subset in X, F is defined as follows [13]:

$$F = \{< x, \mu_F(x) > | x \in X\}$$

where $\mu_F(x)$ is the degree of membership of x in F in the unity interval.

$$\mu_F: X \to [0,1].$$

A fuzzy set is convex if and only if:

$$\mu_F(\lambda x_1 + (1-\lambda)x_2) \geq \min(\mu_F(x_1), \mu_F(x_2))$$

for all $x_1, x_2 \in X$ and $\lambda \in [0,1]$.

A fuzzy set F is normalized if $\sup(\mu_F) = 1$

A fuzzy number is a convex normalized fuzzy set of the real line R^1 whose membership function is piecewise continuous.

A fuzzy number N is called positive, denoted by $N > 0$, if its membership function $\mu_N(x) = 0$ for all $x < 0$

In order to achieve the compromise between simple representation and complexity [14], [15], the most used notation is based on positive trapezoidal fuzzy numbers, where a positive trapezoidal fuzzy number $\tilde{A}$ is denoted by 4-tuple (a_1, a_2, a_3, a_4), (fig. 1) as follows:

$$\mu_{\tilde{A}}(x) = \begin{cases} \dfrac{x-a_1}{a_2-a_1}, & a_1 \leq x \leq a_2 \\ 1, & a_2 \leq x \leq a_3 \\ \dfrac{x-a_4}{a_3-a_4}, & a_3 \leq x \leq a_4 \\ 0, & otherwise \end{cases}$$

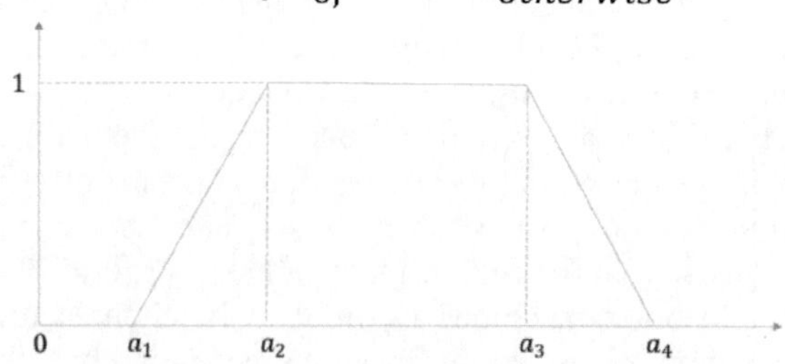

Figure 1. Trapezoidal fuzzy number.

If $a_2 = a_3$ then the trapezoidal fuzzy number is called a triangular fuzzy number. If $a_1 = a_2 = a_3 = a_4$ then we have a classical crisp number.

For any two positive trapezoidal fuzzy numbers $\tilde{A}(a_1, a_2, a_3, a_4)$ and $\tilde{B}(b_1, b_2, b_3, b_4)$ the fuzzy addition, subtraction, multiplication, inverse and division are respectively defined as follows:

$$\tilde{A} \oplus \tilde{B} = (a_1 + b_1, a_2 + b_2, a_3 + b_3, a_4 + b_4)$$

$$\tilde{A} \ominus \tilde{B} = (a_1 - b_4, a_2 - b_3, a_3 - b_2, a_4 - b_1)$$

$$\tilde{A} \otimes \tilde{B} = (a_1 * b_1, a_2 * b_2, a_3 * b_3, a_4 * b_4)$$

$$1 \oslash \tilde{B} = (1/b_4, 1/b_3, 1/b_2, 1/b_1)$$

$$\tilde{A} \oslash \tilde{B} = (a_1/b_4, a_2/b_3, a_3/b_2, a_4/b_1)$$

The distance between two trapezoidal fuzzy numbers $\tilde{A}(a_1, a_2, a_3, a_4)$ and $\tilde{B}(b_1, b_2, b_3, b_4)$ can be computed as follows:

$$D(\tilde{A},\tilde{B}) = \sqrt{\frac{1}{4}[(a_1-b_1)^2+(a_2-b_2)^2+(a_3-b_3)^2+(a_4-b_4)^2]}$$

The distance between two triangular fuzzy numbers $\tilde{A}(a_1,a_2,a_3)$ and $\tilde{B}(b_1,b_2,b_3)$ can be computed as follows:

$$D(\tilde{A},\tilde{B}) = \sqrt{\frac{1}{3}[(a_1-b_1)^2+(a_2-b_2)^2+(a_3-b_3)^2]}$$

For real numbers, the distance between fuzzy numbers becomes the Euclidean distance. Hence, two fuzzy numbers are identical if and only if $D(\tilde{A},\tilde{B}) = 0$. Let $\tilde{A},\tilde{B}$ and $\tilde{C}$ three fuzzy numbers, then $\tilde{A}$ is closer to $\tilde{B}$ than to $\tilde{C}$ if $D(\tilde{A},\tilde{B}) \leq D(\tilde{A},\tilde{C})$.

4. TOPSIS

TOPSIS was proposed in [6] and adopted to the fuzzy context in [7]. The main idea, is that the furthest alternative from the worst solution should be ranked first. This can be calculated by the closeness coefficient for each alternative CC_i, as follows:

$$CC_i = \frac{D_i^-}{D_i^- + D_i^+}, \tag{1}$$

where D_i^- is the distance between the alternative assessment and the worst solution (assessment) and D_i^+ the distance between the alternative assessment and the ideal solution (assessment).
The problem can be formulated as follows:

A set of K Decision Makers (End Users) $E = (E_1, E_2, \dots, E_K)$, are to evaluate a set of m alternatives (smart grids or parts of smart grids) $A = (A_1, A_2, \dots, A_m)$, according to n criteria $C = (C_1, C_2, \dots, C_n)$. The set of performance of A_i with accordance to C_j, $X = (\tilde{x}_{ij}, i = 1, \dots, m; j = 1, \dots, n)$. And criteria impotencies $W = (\tilde{w}_j; j = 1, \dots, n)$
Computing with words introduced by [16] helps meeting more conveniently with human reasoning. Hence, each end-user will make his decision, about criteria importance $\tilde{w}_j$ (Table. 1) and alternatives $\tilde{x}_{ij}$ (Table. 1), by using linguistic variables.

According to the classical approach in TOPSIS, to aggregate p fuzzy numbers $(\tilde{F}_1, \tilde{F}_2, \dots, \tilde{F}_p)$, such that $\tilde{F}_q = (F_{q1}, F_{q2}, F_{q3}, F_{q4})$ where $q = 1, \dots, p$ is $\tilde{F} = (f_1, f_2, f_3, f_4)$ where $f_1 = \min_p\{F_{q1}\}$, $f_2 = \frac{1}{p}\sum_{q=1}^{p} F_{q2}$, $f_3 = \frac{1}{p}\sum_{q=1}^{p} F_{q3}$ and $f_4 = \max_p\{F_{q4}\}$. The main problematic with the classical TOPSIS aggregation is its sensitivity to extrema values, suppose aggregating the three following fuzzy numbers: (1,2,3,4), (7,8,9,10) and (7,8,9,10), the result according to the classical TOPSIS method is (1,6,7,10), which is hugely influenced by the first tuple of the first fuzzy number. Subsequently, we adopt the average value as used for the second and third tuple in the classical TOPSIS method. In the previous example, the result would be (5,6,7,8), that represent conveniently the three aggregated fuzzy numbers.

The method can be summarized as follows:
Step 1: DMs assess criteria importance by linguistic variables (Table 1)
Step 2: DMs assess alternatives with respect to each criterion (Table 1)
Step 3: convert the linguistic variables to trapezoidal/triangular fuzzy numbers (Table 1).
Step 4: construct the weighted fuzzy decisions: Having the individual assessments for each alternative $\tilde{x}_{ij}^K$, the collective assessment for each alternative is : $\tilde{x}_{ij} = \frac{1}{K}[\tilde{x}_{ij}^1 \oplus \tilde{x}_{ij}^2 \oplus \dots \oplus \tilde{x}_{ij}^K]$. And having the individual assessments for weight $\tilde{w}_j^K$, the collective assessment for weights is: $\tilde{w}_j = \frac{1}{K}[\tilde{w}_j^1 \oplus \tilde{w}_j^2 \oplus \dots \oplus \tilde{w}_j^K]$. Hence the weighted fuzzy decision is: $\tilde{G}_{ij} = \tilde{x}_{ij} \otimes \tilde{w}_j$

IOP Conf. Series: Materials Science and Engineering **353** (2018) 012019 doi:10.1088/1757-899X/353/1/012019

$$G = \begin{bmatrix} \tilde{x}_{11}\otimes\tilde{w}_1 & \cdots & \tilde{x}_{m1}\otimes\tilde{w}_1 \\ \vdots & \ddots & \vdots \\ \tilde{x}_{1n}\otimes\tilde{w}_n & \cdots & \tilde{x}_{mn}\otimes\tilde{w}_n \end{bmatrix} = \begin{bmatrix} \tilde{G}_{11} & \cdots & \tilde{G}_{m1} \\ \vdots & \ddots & \vdots \\ \tilde{G}_{1n} & \cdots & \tilde{G}_{mn} \end{bmatrix}$$

Since all fuzzy variables tuples are in the unity interval, there is no need to the normalizing step required in classical TOPSIS

Step 5: determine the fuzzy positive ideal solution (FPIS, $\tilde{A}^+$) and the fuzzy negative ideal solution (FNIS, $\tilde{A}^-$). In the example (Section 5) : $\tilde{A}^+ = (0.7,0.9,0.9)$ and $\tilde{A}^- = (0.1,0.1,0.3)$

Step 6: compute the distance to each candidate

$$D_i^- = \sum_{j=1}^{n} D(\tilde{G}_{ij}, \tilde{A}_j^-) \quad i = 1, \dots, m$$

$$D_i^+ = \sum_{j=1}^{n} D(\tilde{G}_{ij}, \tilde{A}_j^+) \quad i = 1, \dots, m$$

Step 7: compute the closeness coefficients (Eq. 1)

Step 8: rank alternatives according to the closeness coefficients

Table 1. Criteria importance / Alternative evaluation.

Linguistic variable	Fuzzy number
Very low	(0.1,0.1,0.3)
Low	(0.1,0.3,0.5)
Medium	(0.2,0.5,0.7)
High	(0.6,0.7,0.9)
Very high	(0.7,0.9,0.9)

5. Illustrative example and comparison

Authors in [9], proposed a scenario with two users. The first user, who is rich, do not care about cost and wants a maximum comfort using energy. While the second user, who is a green consumer, is concerned about environmental issues. Both users have all needed information that includes cost energy, carbon tax and greenhouse gas GHG emission in home areas; A_1: Kitchen, A_2: bedrooms, A_3: living rooms and A_4: laundry. The collected information, should give users an idea about how efficient they are compared to their neighbors. Table 2 resumes criteria of evaluation and their impotencies assessment according to both users.

Table 2. Linguistic assessment for criteria.

Criteria	Users		Aggregated fuzzy weights
	User 1	User 2	
C_1: Energy cost	H	VH	(0.6,0.8,0.9)
C_2: Budget	M	H	(0.4,0.6,0.8)
C_3: Urgency	H	M	(0.4,0.6,0.8)
C_4: Comfort level	VH	M	(0.5,0.7,0.8)
C_5: GHG emission	VL	VH	(0.4,0.5,0.6)
C_6: Energy efficiency score	L	VH	(0.4,0.6,0.7)
C_7: Carbon tax	M	H	(0.4,0.6,0.8)
C_8: Occupancy level	H	H	(0.5,0.7,0.9)

Table 3. Linguistic assessment for alternatives.

Criteria	Home areas											
	A_1			A_2			A_3			A_4		
	User 1	User 2	Aggregated decision (10^{-1})	User 1	User 2	Aggregated decision (10^{-1})	User 1	User 2	Aggregated decision (10^{-1})	User 1	User 2	Aggregated decision (10^{-1})
C_1	H	M	(4,6,8)	M	H	(4,6,8)	M	VH	(5,7,8)	L	H	(3,5,7)
C_2	L	VH	(4,6,7)	M	L	(2,4,6)	H	L	(3,5,7)	H	L	(3,5,7)
C_3	M	H	(4,6,8)	M	L	(2,4,6)	M	L	(2,4,6)	VH	M	(5,7,8)
C_4	VH	M	(5,7,8)	VH	M	(5,7,8)	VH	H	(6,8,9)	L	VL	(1,2,4)
C_5	L	H	(3,5,7)	M	H	(4,6,8)	L	VH	(4,6,7)	VL	VH	(4,5,6)
C_6	L	M	(2,4,6)	H	H	(5,7,9)	L	VH	(4,6,7)	VL	H	(3,4,6)
C_7	M	M	(3,5,7)	M	VH	(5,7,8)	L	VL	(1,2,4)	M	H	(4,6,8)
C_8	VH	M	(5,7,8)	L	VL	(1,2,4)	VL	M	(2,3,5)	L	M	(2,4,6)

Hence, Table 4 resumes the ranking of alternative after computing all needed parameters

Table 4. Ranking according to closeness coefficients.

alternatives		A_1	A_2	A_3	A_4
Proposed	CC_i	0,331	0,297	0,277	0,201
	Ranking	1	2	3	4
[9] and [11]	CC_i	0,3587	0,3602	0,3539	0,3742
	Ranking	3	2	4	1

The ranking proposed and the ranking in [9], [11] are completely different. While the proposed ranked the first alternative A_1 as first, the ranking in [9], [11] ranked the 4th alternative A_4 as first. Comparing these two alternatives according to their assessments (Table 3) and the criteria impotencies (Table 4), we observe that A_1 is better than A_4 according to criteriaC_1,C_2, C_4 and C_8. A_4 is better than A_1 according to C_3, C_6 and C_7. While there is no preference for neither alternatives according to criterionC_5. Moreover, criteria C_3, C_6 and C_7,that favor A_4, are less or equivalently important compared to other criteria according to their assessments individually.
Discordances between results in our method and other authors [9], [11], are mainly due to the aggregation method proposed in [9], [11], especially the normalizing step, which is unnecessary in our case, since all variables are chosen in the unity interval.

6. Conclusion

The main objective is to know the energy distribution according to users' preferences in smart grids. Fuzzy logic was used to model these preferences, and a modified version of TOPSIS to rank them. A successful comparison proving the consistency of the proposed approach compared to existing approaches in the literature was made.

References

[1] K. Salameh, R. Chbeir, and H. Camblong, "Multi-objective Cooperative Scheduling: An Application on Smart Grids", Appl. Comput. Inform, 2017.
[2] N. U. Hassan, Y. I. Khalid, C. Yuen,S. Huang, M. A. Pasha, K. L.Wood and S. G. Kerk, "Framework for minimum user participation rate determination to achieve specific demand

response management objectives in residential smart grids", *Int. J. Electr. Power Energy Syst.*, vol. 74, no. Supplement C, pp. 91–103, 2016.

[3] D. Geelen, A. Reinders, and D. Keyson, "Empowering the end-user in smart grids: Recommendations for the design of products and services", *Energy Policy*, vol. 61, no. Supplement C, pp. 151–161,. 2013.

[4] G. P. J. Verbong, S. Beemsterboer, and F. Sengers, "Smart grids or smart users? Involving users in developing a low carbon electricity economy," *Energy Policy*, vol. 52, no. Supplement C, pp. 117–125, 2013.

[5] M. Goulden, B. Bedwell, S. Rennick-Egglestone, T. Rodden, and A. Spence, "Smart grids, smart users? The role of the user in demand side management," *Energy Res. Soc. Sci.*, vol. 2, no. Supplement C, pp. 21–29, 2014.

[6] L. A. Zadeh, "Fuzzy sets," *Inf. Control*, vol. 8, no. 3, pp. 338–353, 1965.

[7] R. E. Bellman and L. A. Zadeh, "Decision-Making in a Fuzzy Environment," *Manag. Sci.*, vol. 17, no. 4, pp. B141–B164, 1970.

[8] H. Zhao and N. Li, "Performance Evaluation for Sustainability of Strong Smart Grid by Using Stochastic AHP and Fuzzy TOPSIS Methods," *Sustainability*, vol. 8, no. 2, p. 129, 2016.

[9] O. A. Sianaki and M. A. S. Masoum, "A fuzzy TOPSIS approach for home energy management in smart grid with considering householders' preferences," in *2013 IEEE PES Innovative Smart Grid Technologies Conference (ISGT)*, pp. 1–6. 2013

[10] X. Wang, L. Shi, and C. Zhou, "An interval multiple attribute decision-making model based on TOPSIS and it's application in smart grid evaluation," in *2011 2nd International Conference on Artificial Intelligence, Management Science and Electronic Commerce (AIMSEC)*, pp. 4948–4951, 2011.

[11] O. A. Sianaki and M. A. S. Masoum, "A multi-agent intelligent decision making support system for home energy management in smart grid: A fuzzy TOPSIS approach," *Multiagent Grid Syst.*, vol. 9, no. 3, pp. 181–195, 2013.

[12] J. Zhu, F. Lauri, A. Koukam, and V. Hilaire, "Scheduling Optimization of Smart Homes Based on Demand Response," in *Artificial Intelligence Applications and Innovations*, pp. 223–236, 2015.

[13] I. Skalna, B. Rębiasz, B. Gaweł, B. Basiura, J. Duda, J. Opiła and T. Pełech-Pilichowski, Advances in Fuzzy Decision Making - Theory and Practice Springer. 2015.

[14] P. Grzegorzewski and E. Mrówka, "Trapezoidal approximations of fuzzy numbers," *Fuzzy Sets Syst.*, vol. 153, no. 1, pp. 115–135, 2005.

[15] P. Grzegorzewski and E. Mrówka, "Trapezoidal approximations of fuzzy numbers—revisited," *Fuzzy Sets Syst.*, vol. 158, no. 7, pp. 757–768, 2007.

[16] L. A. Zadeh, "The concept of a linguistic variable and its application to approximate reasoning—I," *Inf. Sci.*, vol. 8, no. 3, pp. 199–249, 1975.

IOP Conf. Series: Materials Science and Engineering **353** (2018) 012020 doi:10.1088/1757-899X/353/1/012020

XRF and UV-Vis-NIR analyses of medieval wall paintings of al-Qarawiyyin Mosque (Morocco)

I Fikri[1], M El Amraoui[1], M Haddad[1], A S Ettahiri[2], L Bellot-Gurlet[3], C Falguères[4], M Lebon[4], R Nespoulet[4], S Ait Lyazidi[1], L Bejjit[1]

[1]Laboratoire de Spectrométrie des Matériaux et Archéomatériaux (LASMAR), URAC11, Faculté des Sciences, Meknès, Université Moulay Ismail (Maroc)
[2]Institut National des Sciences de l'Archéologie et du Patrimoine (INSAP), Rabat (Maroc)
[3]MONARIS, UMR 8233, Université Pierre et Marie Curie, Paris (France)
[4]Muséum National d'Histoire Naturelle, UMR7194, Paris (France)

Email: mhaddad22@yahoo.fr

Abstract. Medieval wall painting fragments, taken at the medieval Mosque of al-Qarawiyyin in Fez, have been investigated by means of X-ray fluorescence and UV-Vis-NIR diffuse reflectance spectroscopies. The analyses permitted to determine the palette of pigments used by craftsmen of the time. Hematite or red ochre were used to obtain red brown colours, calcite for white, copper-based pigments for blue and blue-grey shades while a mixture of cinnabar, lead-based pigments and hematite was adopted to make red-orange colours. Furthermore, the analysis of mortars (external layer and plaster) on these wall painting samples revealed that they are composed mainly by calcite and sometimes by additional compounds such as quartz and gypsum.

1. Introduction

In the recent decade, there has been a great interest in the use of non-invasive analytical techniques able to answer specific questions about materials used in handcrafting of ancient artworks, for conservation sciences, archaeological studies and art history goals [1-3]. The present work relates to medieval wall paintings, dating back to the 11th-12th centuries, at the al-Qarawiyyin Mosque of Fez. The Mosque al-Qarawiyyin, funded in the 9th century, is the oldest university in the Islamic world. Most studies conducted on Islamic wall paintings deal with those in Spain [2,4-7]; none study concerned those of Morocco.

Numerous wall paintings fragments, originating from al-Qarawiyyin Mosque, are investigated by mean of combining elementary and structural analyses in order to characterize painting and mortar layers. The adopted techniques are X-ray fluorescence and UV-Vis-NIR diffuse reflectance spectroscopies; both are non-invasive, not requiring any material removing from the painting surface. X-ray fluorescence permits elementary analyses identifying chemical elements in inorganic materials [8], while UV-Vis-NIR diffuse reflectance allows structural analyses of pigments [9-13] and colourants with a high degree of probability [14].

The wall painting fragments were collected in the course of a recent excavation in al-Qarawiyyin architectural site. They might originate from two neighbouring houses demolished during the extension of the mosque in 1134, during the Almoravids Dynasty period. Only results relating to six representative fragments are shown.

The work is part of a research project aiming to find more information to differentiate between materials and handcrafting techniques adopted in Mediterranean sites. The results will also enlarge an

IOP Conf. Series: Materials Science and Engineering **353** (2018) 012020 doi:10.1088/1757-899X/353/1/012020

ongoing database on ancient Moroccan wall paintings from the Roman period (Banasa and Volubilis sites) to the medieval Islamic one.

2. Materials and Techniques

2.1. Materials

Six representative wall painting fragments (figure 1), among a large set of pieces dating back to the 11th -12th century period taken at al-Qarawiyyin Mosque, were studied. The samples exhibit different colourful decorative patterns designed on mortar surfaces. They are of irregular sizes varying from 4 cm x 4 cm to 17 cm x 6 cm, with thicknesses in the range 1 to 3 cm.

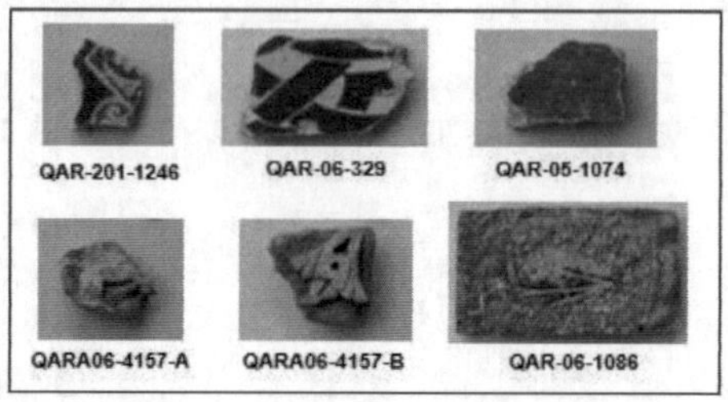

Figure 1. Photographs of the wall painting studied samples

The six fragments are referenced QAR-201-1246, QAR-06-329, QAR-05-1074, QAR-06-1086, QARA06-4157-A and QARA06-4157-B. On the basis of careful macroscopic observations, the two first pieces show dichromatic decorations in white and dark red-brown shades, the third one appears as a dark red-brown monochromatic surface, the fourth fragment presents red-orange and blue-grey areas isolated by thin black strips, while the fifth and sixth ones look like carved mortars with small areas of faint blue pigmentation. The patterns colours are applied on white or yellowish layers of mortars, with very fine aggregates.

2.2. Techniques

All reflectance spectra were recorded on a JASCO V-570 UV-Vis-NIR spectrometer equipped with an integrating sphere; the reference is a BaSO4 tick plaque. All measures are repeated several times upon varying the spotted zone on the fragment surface. Pigments identification was achieved by mean of reflectance spectra databases [15].

XRF elementary analyses were carried out on an ELIO portable energy dispersive spectrometer managed by XG ELIO software, highly sensitive in the full range 1–40 keV. The X-ray source is a tube of Rh-anode operating at maximum (40 kV - 200 μA). The spectra were measured on spots of 0.25 mm2 size area during 300 seconds. The instrument has a resolution lower than 140 eV at line peak Kα= 5.9 keV of Manganese. Chemical elements were identified on the basis of their characteristic X-ray peaks energies [1].

3. Results and discussion

Figure 2 shows the XRF spectra recorded on all fragments. The main elements detected in these samples are calcium (Ca), iron (Fe), copper (Cu), mercury (Hg), sulfur (S); and lead (Pb). To complete this qualitative analysis by a structural one, UV-Vis-NIR spectra were recorded.

3.1. Red-brown and white paintings

Only the three red-brown colour samples referenced QAR-05-1074, QAR-201-1246 and QAR-06-329 have been analysed by optical reflectance spectrometry (figure 3); it was not possible to analyse the other fragments because of their dimensions and their geometrical. All red-brown areas show similar spectra exhibiting their maximum at 740 nm and characterized by a sharp positive slope between 550 and 600 nm. A strong absorption band around 490 nm along with another weak one in the near infrared 840-870 nm side are also observed. These latter bands are attributed to electronic transitions of Fe^{3+} ions [16-19]. This is in accordance with the red-brown's reflectance spectrum first derivative (figure 3 FD) which shows an inflexion point around 585 nm characteristic of iron Fe^{3+}. From these results, it seems that red ochre or hematite (α-Fe_2O_3) had been used to achieve the red-brown colour

IOP Conf. Series: Materials Science and Engineering **353** (2018) 012020 doi:10.1088/1757-899X/353/1/012020

[3,16,20,21]. XRF spectra (figure 2.I-a,b,c) support then this finding; they revealed Iron (Fe) and Calcium (Ca) as principal chemical elements, detected in high content, in the red-brown areas of samples QAR-05-1074, QAR-06-329 and QAR-201-1246. Hematite used to be preferred as a red pigment since antiquity because of its availability, easy preparation and also low cost. Its deep colour gives an exceptional contrast with white backgrounds promoting thus the visual effect, such as the case of the present dichromatic samples [5].

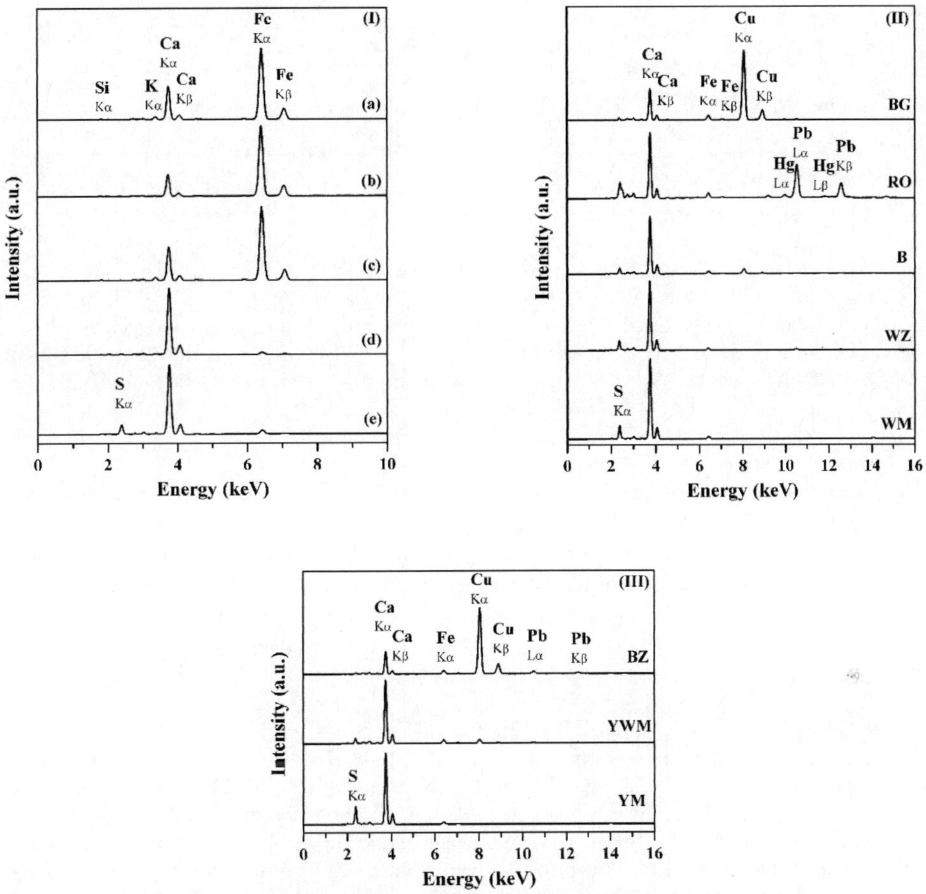

Figure 2. XRF spectra

(I) Dark red-brown areas in samples QAR-05-1074 (a), QAR-201-1246 (b) and QAR-06-329 (c); white zone (d) and white mortar (e) in sample QAR-06-329.
(II) Blue grey (BG), red orange (RO), black (B), white zone (WZ) and white mortar (WM) in sample QAR-06-1086.
(III) Blue zone (BZ), yellowish white (YWM) and white (WM) mortars in sample QARA06-4157-A.

On all reflectance spectra there are three additional absorption bands observed around 1450, 1740 and 1950 nm which can be associated to natural gypsum (calcium sulfate dihydrate $CaSO_4 \cdot 2H_2O$) [18,22], or to carbonate ion CO_3^{2-} resulting from the presence of calcite ($CaCO_3$) in the red-brown areas [18]. The observed high relative intensity of calcium in white zones (figure 2) is indicative of using calcite ($CaCO_3$); calcite used to be adopted to produce white colours [16,23]. Calcite has been also revealed by Raman measurements (results to be published). Calcium is also present in all samples areas; it was likely used either as a principal component to achieve under layer (called intonaco) of paintings or as a shade lightening.

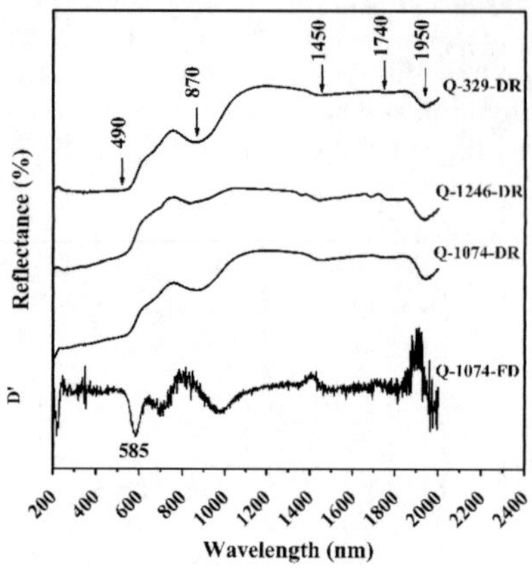

Figure 3. Reflectance spectra of dark red-brown colour areas (DR) in samples QAR-06-329, QAR-201-1246 and QAR-05-1074; diffuse reflectance first derivative spectrum (FD) of QAR-05-1074 (bottom).

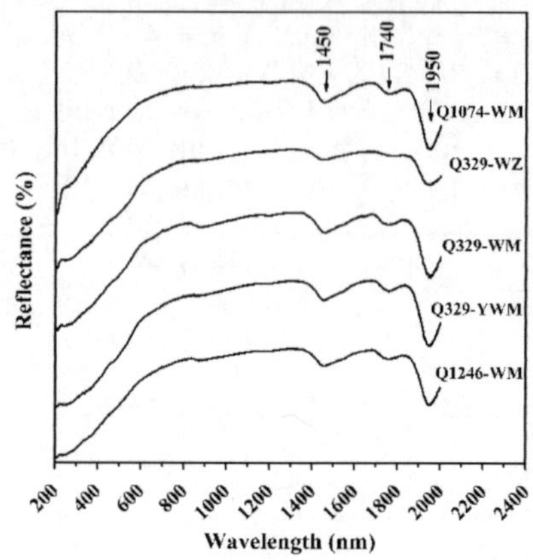

Figure 4. Reflectance spectra of white mortar (WM) in fragment QAR-05-1074; white zone (WZ), white mortar (WM) and yellowish white mortar (YWM) in fragment QAR-06-329 and white mortar (WM) in fragment QAR-201-1246.

3.2. Blue colour painting

XRF measured on the blue colour areas of samples QARA06-4157-A and QARA06-4157-B are shown in (figure 2.III-BZ) The analyses highlight the presence of copper (Cu) in high content. This element points out that a Cu-based pigment was used to achieve the blue colour of the painting, Raman measurements has been used in this case to specify the identity of the Cu-based pigment which is Azurite [$Cu_3(CO_3)_2\,(OH)_2$] (results to be published). Calcium (Ca), iron (Fe) and lead (Pb) are also observed in all blue areas. As underlined above, neither the dimensions nor the geometrical 3D-forms of these fragments permitted UV-Vis-NIR diffuse reflectance measurements on them.

3.3. Polychrome painting

XRF measurements recorded on red-orange areas of sample QAR-06-1086 (figure 2.II-RO) revealed Lead (Pb), mercury (Hg), sulfur (S) and calcium (Ca) as principal chemical elements in this painting. The presence of these chemical elements suggests that a mixture of mercuric sulphide (HgS) and a Pb-based pigment such as minium (Pb_3O_4) was adopted to obtain the red-orange colour [13,17,25]. Iron (Fe) is also present in a low amount, this refers to hematite (α-Fe_2O_3) or red ochre both containing iron (Fe). Therefore, all the pigments, minium, cinnabar or vermilion, hematite and/or red ochre were used to produce the red-orange colour.

Copper (Cu) is the principal chemical element identified by XRF measurements in the blue-grey areas of QAR-06-1086 polychrome sample (figure 2.II-BG). This result points out that a copper-based blue pigment such as Azurite [Cu3 (CO3)2 (OH)2] or Egyptian blue had been used [21]. For black areas on the fragment the XRF spectra didn't indicate any elements which could be related to the black colour. Consequently, black paintings were very likely realized by mean of an organic compounds-based receipt such as carbon black or ivory whose chemical elements cannot be seen in XRF spectra [1, 11].

IOP Conf. Series: Materials Science and Engineering **353** (2018) 012020 doi:10.1088/1757-899X/353/1/012020

3.4. Mortars

XRF spectra measured on mortars layers (figures 2. I-e, 2.II-WM and 2.III-WM) show high amounts of calcium (Ca) with relatively lower amounts of sulfur (S). This result suggests that calcite ($CaCO_3$) was the principal component in the present mortars [17,26]. This is consistent with UV-Vis-NIR reflectance measurements (figure 4) revealing bands attributable to calcium sulphate dehydrate ($CaSO_4 . 2H_2O$). Quartz is also highlighted in these materials indicating thus that sand was likely used with white lime (calcite) to produce layers of plasters.

4. Conclusion

A set of six representative wall painting fragments, chosen among a large series of pieces sampled at the medieval al-Qarawiyyin Mosque of Fez, have been characterized by mean of crossing elementary XRF analyses and structural UV-Vis-NIR diffuse reflectance spectrometry techniques. The adopted colouring materials in these paintings have been identified and characterized.

The measurements revealed that: i) red-orange shades are mainly lead-based pigments types with additional cinnabar and traces of hematite or red ochre, ii) red-brown colours had been achieved by hematite or red ochre mixed with some amounts of black pigments, iii) blue and blue-grey colours are copper-based pigments, while vi) calcite used to be adopted to make white colour of painting layers. The investigations showed also that mortars of wall paintings are composed principally of calcite with minor amount of gypsum.

The identified palette of painting materials used to be adopted by craftsmen of the period seems likely arranged by local and imported pigments. These results are of great importance while choosing painting materials for future restoring operations.

Acknowledgments

This study was supported by Moroccan CNRST and French CNRS, and by ColorHOM project.

References:

[1] Kriznar A, Munoz M D V, De la Paz F, Respaldiza M A and Vega M 2014 Non-destructive XRF analysis of selected Flemish panel paintings in the Fine Arts Museum of Seville, *J. of the Institute of Conservation* **37 (2)** 136-51.

[2] Dominguez-Vidal A, De la Torre-López M J, Campos-Suñol M J, Rubio-Domene R and Ayora-Cañada M J 2014 Decorated plasterwork in the Alhambra investigated by Raman spectroscopy: comparative field and laboratory study, *J. Raman Spectrosc.* **45** 1006–12.

[3] Bersani D, Berzioli M, Caglio S, Casoli A, Lottici P P, Medeghini L, Poldi G and Zannini P 2014 An integrated multi-analytical approach to the study of the dome wall paintings by Correggio in Parma cathedral, *Microchem J.* **114** 80–88.

[4] Garofano I, Duran A, Perez-Rodriguez J L and Robador M D 2011 Natural Earth Pigments from Roman and Arabic Wall Paintings Revealed by Spectroscopic Techniques, *Spectrosc. Lett.* **44** 560-65.

[5] Duran A, Perez-Rodriguez J L, Jimenez de Haro M C, Franquelo M L and Robador M D 2011 Analytical study of Roman and Arabic wall paintings in the Patio De Banderas of Reales Alcazares' Palace using non-destructive XRD/XRF and complementary techniques, *J. Archaeol. Sci.* **38** 2366-77.

[6] Robador M D, DE Viguerie L, Pérez-Rodríguez J L, Rousselière H, Walter P and Castaing J 2016 The structure and chemical composition of wall paintings from Islamic and Christian times in the Seville alcazar, *Archaeometry.* **58** 255–70.

[7] Garofano I, Robador M D and Duran A 2014 Materials characteristics of roman and 5evill mortars and stuccoes from the patio de banderas in the real alcazar of Seville (spain), *Archaeometry.* **56** 541-61.

[8] El Amraoui M, Haddad M, Bejjit L, Ait Lyazidi S, and Lakhal R 2017 On-site XRF characterization of archaeological materials in CERA center of Rissani (Morocco), *IOP Conf. Ser.: Mater. Sci. Eng.* **186** 012029.

[9] Dupuis G, Elias M and Simonot L 2002 Pigment Identification by Fiber-Optics Diffuse Reflectance Spectroscopy, *Appl. Spectrosc.* **56 (10)** 1329–36.

[10] Hartwig M 2013 *Visual and Archaeometric Analysis of the Paintings.* The tomb chapel of

Menna (TT 69). The art, culture, and science of painting in an Egyptian tomb, New York, The American University in Cairo Press. Chapter 6, 133-161.
[11] Edreira M C, Feliu M J, Fernández-Lorenzo C and Martı n J 2001 Roman wall paintings characterization from Cripta del Museo and Alcazaba in Mérida (Spain): chromatic, energy dispersive X-ray fluresence spectroscopic, X-ray diffraction and Fourier transform infrared spectroscopic analysis, *Analytic Chimi Act.* **434** 331–45
[12] Aceto M, Agostino A, Fenoglio G, Idone A, Gulmini M, Picollo M, Ricciardi P and Delaney J K 2014 Characterisation of colorants on illuminated manuscripts by portable fibre optic UV-visible-NIR reflectance spectrophotometry, *Anal. Methods*. **6** 1488.
[13] Garofano I, Perez-Rodrigueza J L, Robador M D, Duran A 2016 An innovative combination of non-invasive UV–Visible-FORS, XRD and XRF techniques to study Roman wall paintings from Seville, Spain, *J. Cult. Herit.* **22** 1028-39.
[14] Oltrogge D 2008 *The Use of VIS Spectroscopy in Non-Destructive Paint Analysis*, Research Project Painting Techniques of Impressionism and Postimpressionism, Online-Edition www.museenkoeln.de/impressionismus, Köln.
[15] Web site database: http://fors.ifac.cnr.it/.
[16] Cheilakou E, Troullinos M, Koui M 2014 A identification of pigments on Byzantine wall paintings from Crete (14th century AD) using non-invasive Fiber Optics Diffuse Reflectance Spectroscopy (FORS), *J. Archaeol. Sci.* **41** 541-55.
[17] Bakiler M, Kı rmı zı B, Öztürk Ö O, Hanyalı Ö B, Dağ E, Çağlar E and Köroğlu G 2016 Material characterization of the Late Roman wall painting samples from Sinop Balatlar Church Complex in the black sea region of Turkey, *Microchem J.* **126** 263–73.
[18] Clark R N 1999 Spectroscopy of rocks and Minerals, and principes of spectroscopy, *Manual of Remote Sensing* **3(3-58)** 2-2.
[19] Pailhe N, Wattiaux A, Gaudon M and Demourgues A 2008 Correlation between structural features and vis–NIR spectra of a-Fe_2O_3 hematite and AFe_2O_4 spinel oxides (A = Mg, Zn) *J. Solid. State. Chem.* **181** 1040–47.
[20] Garofano I 2015 *Caracterización físico-química de morteros de revestimiento y pinturas murales, de época romana e islámica de la excavación arqueológica del Patio de Banderas, y cerámicas vidriadas de los siglos XI-XV, del Real Alcázar de Sevilla.*(Doctoral dissertation). Universidad de Sevilla, Sevilla.
[21] Edreira M C, Feliu M J, Fernandez-Lorenzo C and Martı n J 2003 Spectroscopic analysis of roman wall paintings from Casa del Mitreo in Emerita Augusta, Me´rida, Spain, *Talanta* **59** 1117-39.
[22] Moreira L C J, Teixeira A d S and Galvão L S 2014 Salinization of Brazilian Alluvial Soils and the Spectral Effects of Gypsum, *Remote Sens.* ***6***, 2647-63.
[23] Baraldi P, Bracci S, Cristoferi E, Fiorentino S, Vandini M and Venturi E 2016 Pigment characterization of drawings and painted layers under 5th–7th centuries wall mosaics from Ravenna (Italy), *J. Cult. Herit.* **21** 802–08.
[24] El Bakkali A, Lamhasni T, Ait Lyazidi S, Haddad M, Rosi F, Miliani C, Sánchez-Cortésc S, El Rhaiti M 2014 Assessment of a multi-technical non-invasive approach for the typology of inks, dyes and pigments in two 19th century's ancient manuscripts of Morocco, *Vib. Spectrosc*. **74** 47–56.
[25] Avlonitou L 2016 Pigments and colors: An inside look at the painted decoration of the Macedonian funerary monuments, *J. Archaeol. Sci*. Reports **7** 668–78.
[26] Syta O, Rozum K, Choińska M, Zielińska D, Żukowska G Z, Kijowska A and Wagner B 2014 Analytical procedure for characterization of medieval wall-paintings by X-ray fluorescence spectrometry, laser ablation inductively coupled plasma mass spectrometry and Raman spectroscopy, *Spectrochim. Acta. Part B* **101** 140–48.

IOP Conf. Series: Materials Science and Engineering **353** (2018) 012021 doi:10.1088/1757-899X/353/1/012021

Regulation of the Output Voltage of an Inverter in Case of Load Variation

Omar DIOURI, Fatima ERRAHIMI, Najia ES-SBAI

Laboratory of Renewable Energies and Intelligent Systems, Faculty of Sciences and Technology, Sidi Mohamed Ben Abdellah University, Fez, Morocco

Email: omar.diouri1@usmba.ac.ma

Abstract. In a DC/AC photovoltaic application, the stability of the output voltage of the inverter plays a very important role in the electrical systems. Such a photovoltaic system is constituted by an inverter, which makes it possible to convert the continuous energy to the alternative energy used in systems which operate under a voltage of 230V. The output of this inverter can be connected to a single load or more, at which time a second load is added in parallel with the first load. In this case, it proves a voltage drop at the output of the inverter. This problem influences the proper functioning of the electrical loads. Therefore, our contribution is to give a solution to this by compensating this voltage drop using a boost converter at the input of the inverter. This boost converter will play the role of the compensator that will provide the necessary voltage to the inverter in order to increase the voltage across the loads. But the use of this boost without controlling it is not enough because it generates a voltage that depends on the duty cycle of the control signal. To stabilize the output voltage of the inverter, we used a Proportional, Integral, and Derivative control (PID), which makes it possible to generate the necessary control signal for the voltage boost in order to have a good regulation of the output voltage of the inverter. Finally, we have solved the problem of the voltage drop even though there is loads variation.

1. Introduction

The rising energy demand and the negative effect of fossil fuel require the increase of electricity generation via renewable energy sources [1]. Wind energy, fuel cells and solar PV are used in order to meet the necessary demands of loads and inject this green electricity into the grid. This can be named a microgrid [2]. In PV applications, there are many structures used to interface with the loads or the grid. To realize this interface, power electronics like a DC/DC converter and a DC/AC inverter are used. In literature, many topologies of power electronic systems are used in order to increase the efficiency by minimizing the number of switches, reducing the Total Harmonic Distortion (THD) and regulating the voltage and current to be injected in the grid [3]. On the other hand, the control of these converters offers an efficient tool in regulating some parameters or adjusting the power injection [4]. The majority of research focuses on how to extract the maximum power using "Perturb and Observe" algorithm [5] and fuzzy logic control [6]. It also focuses on determining the state of charge of the battery using a Kalman filter [7] and Neural Networks [8]. However, some applications require an output voltage of 230V in the inverter control. In case of load variation, the DC voltage of the boost converter should be kept near the desired value and the voltage drop phenomenon must be avoided. Therefore, it is necessary to use a control scheme in order to regulate the AC output voltage of the system in a way that guarantees stability. DC/DC boost converter uses the proportional integral (PI) controllers to regulate a DC bus voltage [4-9]. However, the

IOP Conf. Series: Materials Science and Engineering 353 (2018) 012021 doi:10.1088/1757-899X/353/1/012021

identification of PID gains depends on the whole system parameters, the input voltage and the load variations. In this paper, we assume that there is no variation of input voltage and we implement a PID control regarding load variations at AC output inverter. The paper is organized as follow: in section 2 we describe our system with determine the parameters of each bloc. After that, in section 3, we give simulation results of two different circuits for this system, the first circuit is the inverter without boost converter in order to show the influence of the load variation on the inverter output voltage, the second circuit is the boost converter connected to the inverter using PID control for regulating and solving the problem of voltage drop in the output in case of load variation. The Last section is conclusion.

2. System Description

2.1. DC Voltage Source

The PV panel produces direct DC current from sunlight. This current can be used in electrical equipment or to recharge a battery. However, this green electricity generated depends on solar irradiation and temperature, which is reflected in a variation of the voltage and power in general. In order to study only the variation effect of the loads on the stability of voltage, we replaced the PV panel with a DC Voltage source to fix the input voltage of the converter stage. In a first step, without Boost converter, the value of the DC Voltage source used is 326V to get 230V which is the Root Mean Square (RMS) value of the voltage. In the second step, with Boost converter, the value of the DC Voltage source is 100V. It is less than the previous step because we added a Boost converter to the system in order to increase the input voltage of the inverter in case of a voltage drop.

2.2. Boost Converter

In literature, there are different types of DC/DC converters that can transform a certain level of voltage to another. They can be voltage boosters or step down converters. In this work, we used a DC/DC Boost converter as shown in Figure 1. This model requires finding the appropriate value of the inductor and the capacitor to have the desired voltage at the output of this boost converter.

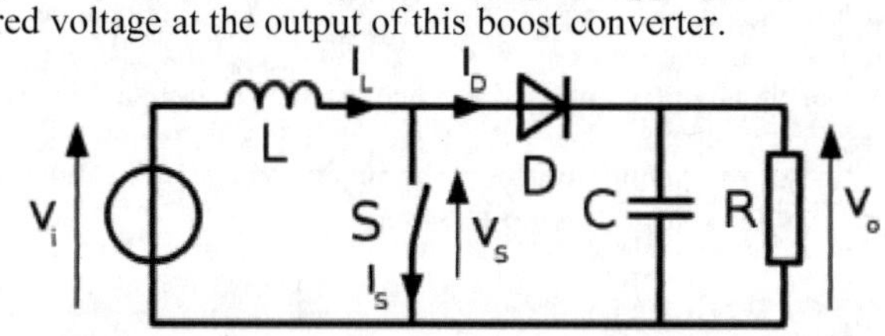

Figure 1. Model circuit of DC/DC Boost converter

To determine the value of the inductor and capacitor, the following equations must be used [5].

$$V_0 = \frac{V_i}{1-\alpha} \quad (1)$$

$$I_0 = I_L(1-\alpha) \quad (2)$$

Where, α is the duty cycle, V_0 and I_0 are the output voltage and output current respectively, V_i and I_L are the input voltage and input current respectively.
The value of inductor and capacitor is given by equations (3) and (4) respectively.

$$L = \frac{\alpha V_i}{F \Delta I_L} \quad (3)$$

$$C = \frac{\alpha I_L}{F \Delta V_0} \quad (4)$$

F is the switching frequency, ΔV_0 is the ripple of output voltage, ΔI_L is the ripple of inductor current.

2.3. Inverter

The main objective of an inverter is to convert DC current to AC current [3]. As shown in figure 2, the inverter used in this paper is composed of four switches that are MOSFET transistors to obtain an H-Bridge inverter. The main objective of this part is the conversion of voltage from DC to AC.

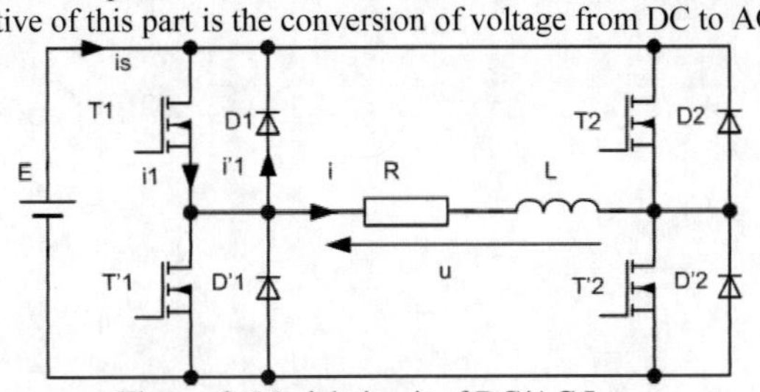

Figure 2. Model circuit of DC/AC Inverter

2.4. Low-pass Filter

In this paper we used a passive low-pass filter LC which is composed of an inductor and capacitor. It is placed between the inverter and the loads. This filter allows us to obtain a purely sinusoidal voltage, in order to use it by different loads.

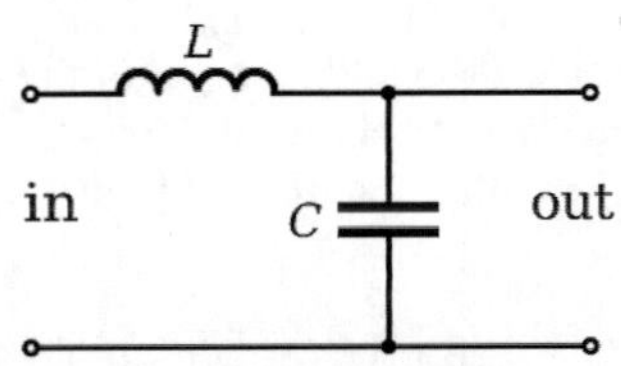

Figure 3. LC filter

The value of inductor and capacitor is calculated by equations (5) and (6) as follow:

$$L < \frac{0.03U_{inv}}{2\pi f I_{Lmax}} \tag{5}$$

$$C < \frac{1}{(2\pi f_c)^2 L} \tag{6}$$

Where, U_{inv} is the maximum value of the inverter output voltage, I_{Lmax} is the maximum value of the inductor current, f is the frequency which is 50Hz and finally f_c is the cutoff frequency.

2.5. Loads

The final part is loads that consume the electrical power. In this work, we used different values of-simple resistive loads to examine the inverter output voltage in case of load variation. At first, we used a single charge fed directly to the output of the inverter. After a while, we added a second load in parallel in order to have a growth of the electrical consumption. After that, there is also the addition of a third load in parallel.

3. Simulation results and discussions

We simulated this work in two parts. The first is the variation loads at the output of inverter without using a boost converter which means without regulation. The second is made by adding a stage of boost converter at the input of the inverter to ensure a stable voltage. Simulation is done using Simulink.

3.1. The effect of the load variation on the inverter output voltage

The simulation circuit is illustrated in the figure 4.

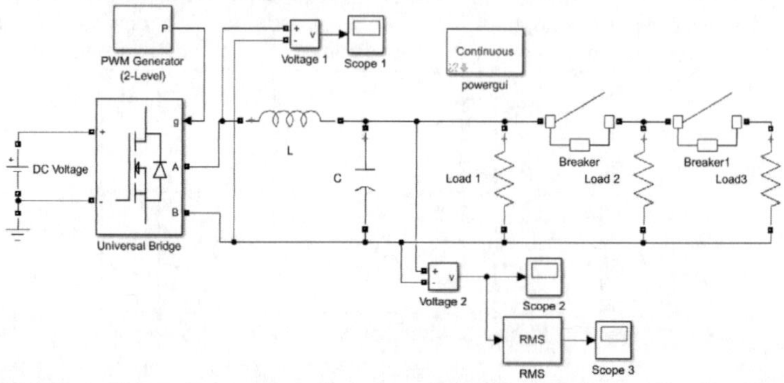

Figure 4. Simulation of the output voltage inverter with different value of load, without regulation

We obtained directly at the output voltage of the inverter (Scope1), an alternative voltage as shown in figure 5. After filtering this voltage by the LC filter and making some variation of load at different times, the phenomenon observed is the voltage drop at the output of the inverter (Scope2) as shown in figure 6(a). In figure 6(b), a zoom at the moment t = 1s shows the effect of the addition of a second load. Figure 7 illustrates the RMS value of the output voltage at different times with load variation.

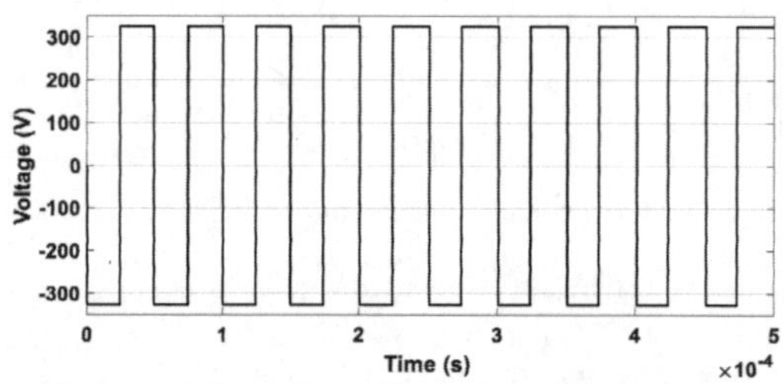

Figure 5. Output voltage inverter before filter

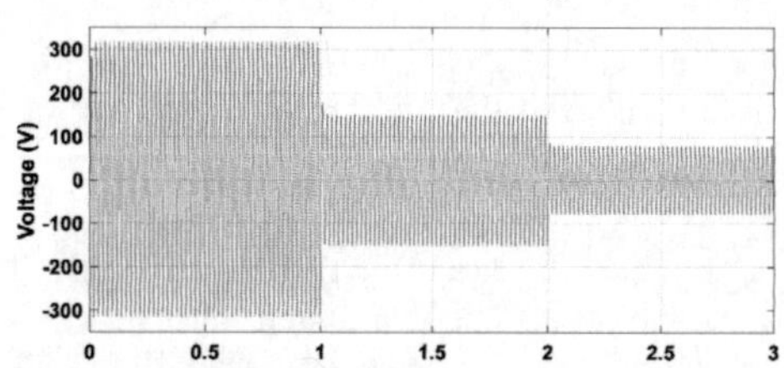

Figure 6a. Output voltage of inverter with load variation without regulation

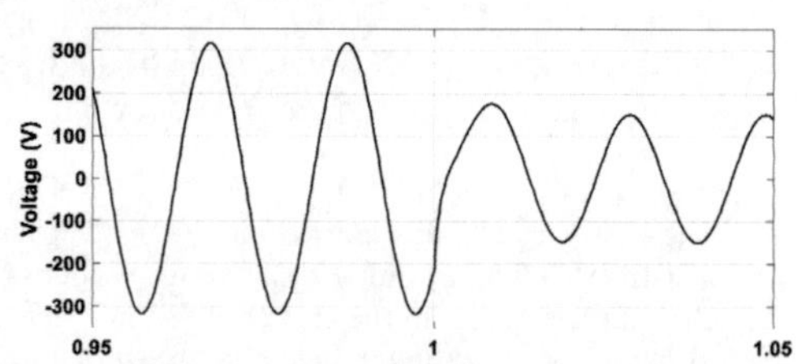

Figure 6b. Zoom at t=1s on the output voltage of inverter

IOP Conf. Series: Materials Science and Engineering **353** (2018) 012021 doi:10.1088/1757-899X/353/1/012021

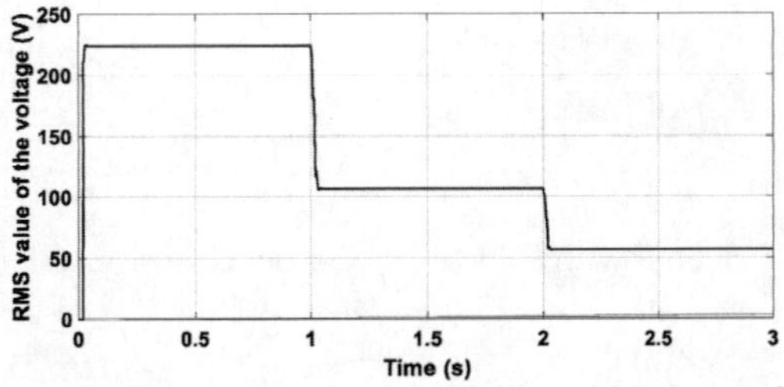

Figure 7. RMS value of output voltage without regulation

3.2. Regulation of the output voltage in case of variation load

The simulation circuit is illustrated in the figure 8.

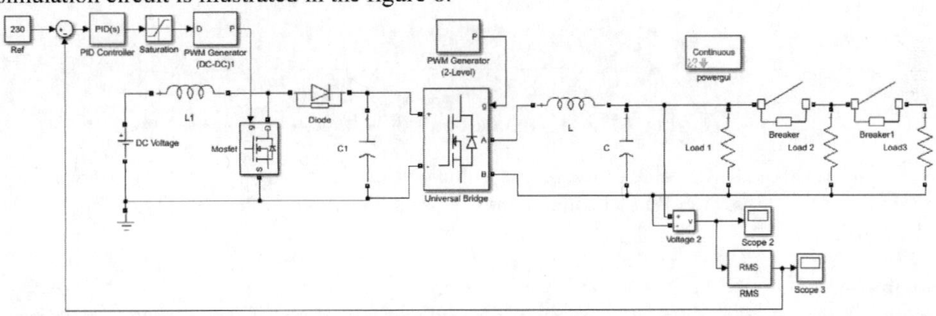

Figure 8. Simulation of the output voltage inverter with different value of load, with PID Control

The problem of voltage drop in case of load variation is solved by using the PID control for a boost converter in order to regulate the output voltage of the inverter if there is a phenomenon of voltage drop. As shown in figure 8, the reference RMS voltage is 230V and it is compared with the RMS value of the output voltage inverter. In case of voltage drop following the addition of a new load, the PID controller plays a very important role in adapting the duty cycle in order to regulate the voltage at the output of the inverter. To simulate this circuit, we used the following parameters as illustrated in table 1.

Table 1. Parameters of circuit elements.

Parameter	Value	Parameter	Value	Parameter	Value
VDC	100 V	**F_{PWM}**	20 KHz	**Kp**	0.0013
Vref (RMS)	230 V	**f**	50 Hz	**Ki**	0.0318
L1 (Boost)	9.90e-4 H	**L (filter)**	2.97e-3 H	**Kd**	3.0479e-15
C1 (Boost)	3.33e-5 F	**C (filter)**	1.18e-6 F	**Load1**	50 W
Load2	100 W	**Load3**	150 W		

As shown in figure 9a and figure 9b, we solved the problem of the voltage drop following an addition of different loads. At t = 1s, the addition of a second load causes a decrease in voltage. The PID controller can generate the appropriate duty cycle to provide the voltage needed to obtain an RMS value of 230V.

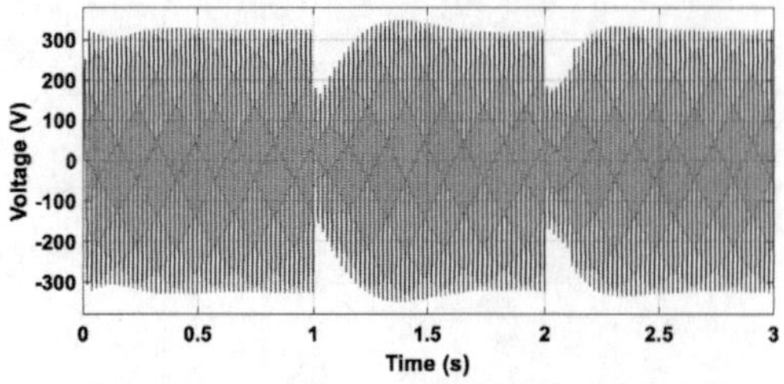

Figure 9a. Output voltage of inverter with load variation with regulation

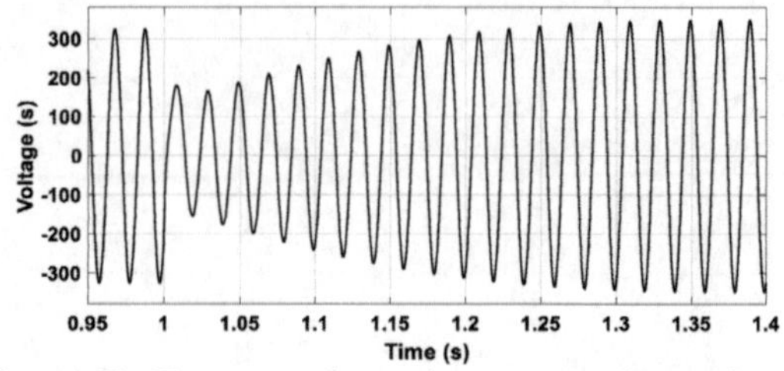

Figure 9b. Zoom at t=1s on the output voltage of inverter

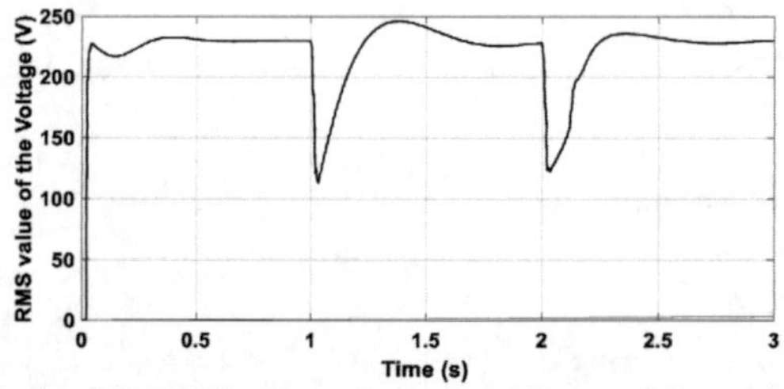

Figure 10. RMS value of output voltage with regulation

Figure 10 illustrates the RMS value of the output voltage with PID control. As shown, despite the response time that exceeds 600 ms, the PID controller achieves our goal, which is the stability of the voltage on a value of 230V.

4. Conclusion

The model of the inverter with load variation is simulated and tested using Simulink platform. The PID control is used as a corrector of voltage drop in the output voltage of the inverter. To meet international requirements and standards, the voltage must be 230V for the electrical devices. This has pushed us to realize a solution that is efficient and simple to have stability in the voltage of the inverter even if there is a change of loads. To improve the system, we plan to study our system in the case of inductive loads.

References

[1] P.Monica, M.Kowsalya, P.C.Tejaswi, "Load sharing control of parallel operated single phase inverters", Energy Procedia, Volume 117, June 2017, Pages 600-606, https://doi.org/10.1016/j.egypro.2017.05.156

[2] Lasseter, Robert H. Microgrids. Power Engineering Society Winter Meeting, IEEE; 27-31 January 2002. Vol. 1. p. 305-8

[3] Omar DIOURI, Ahmed GAGA, Najia ES-SBAI, Fatima ERRAHIMI, A. (2015). Design and simulation of a novel cascaded transformer multilevel inverter topology for photovoltaic system, In 3rd International Renewable and Sustainable Energy Conference (IRSEC) 10.1109/IRSEC.2015.7454957

[4] George C. Konstantopoulos, Antonio T. Alexandridis, Non-linear voltage regulator design for DC/DC boost converters used in photovoltaic applications: analysis and experimental results, The Institution of Engineering and Technology 2013, IET Renew. Power Gener., 2013, Vol. 7, Iss. 3, pp. 296–308

[5] Motahhir, S., El Ghzizal, A., Sebti, S., & Derouich, A. (2015). Proposal and implementation of a novel perturb and observe algorithm using embedded software. In 3rd International Renewable and Sustainable Energy Conference (IRSEC), (pp. 1–5). IEEE.

[6] Di Piazza, M.C., Pucci, M., Ragusa, A., Vitale, G.: 'Fuzzified PI voltag control for boost converters in multi-string PV Plants'. Proc. 34th IEEE Annual Conf. Industrial Electronics IECON 2008,

Orlando, FL, 10–13 November 2008, pp. 2338–2345

[7] Ahmed Gaga, Hicham Benssassi, Fatima Errahimi, Najia Es-Sbai, Battery state of charge estimation using an adaptive unscented kalman filter for photovoltaic applications, International Review of Automatic Control (IREACO), Vol 10, No 4 (2017)

[8] Youssef Cheddadi, Omar Diouri, Ahmed Gaga, Fatima Errahimi, Najia Es-Sbai. "Design and Simulation of an Accurate Neural Network State-of-Charge Estimator for Lithium Ion Battery Pack", International Review of Automatic Control (IREACO), Vol. 10, N. 2 (2017)

[9] Amin, M.N., Elshaer, M.A., Mohammed, O.A.: 'DC bus voltage control for PV sources in a DC distribution system infrastructure'. Proc. 2010 IEEE Power and Energy Society General Meeting, Minneapolis, MN, 25–29 July 2010, pp. 1–5

IOP Conf. Series: Materials Science and Engineering **353** (2018) 012022 doi:10.1088/1757-899X/353/1/012022

Pubface: Celebrity face identification based on deep learning

H Ouanan[1*], M Ouanan[1] and B Aksasse[1]

[1]Dept. of Computer Science, ASIA Team, M2I Laboratory, Faculty of science and techniques, Errachidia, Morocco

Email: *ham.ouanan@gmail.com

Abstract. In this paper, we describe a new real time application called PubFace, which allows to recognize celebrities in public spaces by employs a new pose invariant face recognition deep neural network algorithm with an extremely low error rate. To build this application, we make the following contributions: firstly, we build a novel dataset with over five million faces labelled. Secondly, we fine tuning the deep convolutional neural network (CNN) VGG-16 architecture on our new dataset that we have built. Finally, we deploy this model on the Raspberry Pi 3 model B using the OpenCv dnn module (OpenCV 3.3).

1. Introduction

The human face plays an important role in our social interaction, carrying information about people's identity. Due to recent progress in performance, the use of facial recognition technology has grown significantly in the past several years due to its potential for a wide variety of applications in both commercial applications and security. Compared to other biometric technology, face recognition techniques does not require the collaboration of the subject. It enables faces to identify a person in real-time.

Recently face recognition systems was known extraordinary breakthrough. Much of this progress can be attributed to the success of deep learning based techniques, which are actively being used in many areas where traditional algorithmic solutions don't work well or don't work at all. These techniques perform an assumption that by collecting massive training sets, deep networks will have sufficient examples of both inter-subject and intra-subject appearance variations. From these variations, artificial neural networks can learn to generate discriminative features, which amplify subject identity and suppress other, confounding appearance variations. So, to train face recognition systems based on deep CNNs, very large training sets are needed with millions of labeled images. Recently, the number of face images has been growing exponentially on social network such as Facebook and Twitter. As an example, the director of Facebook AI Research Yann LeCun has said, "almost 1 billion new photos were uploaded each day on Facebook in 2016". However, large training datasets are not publicly available and very difficult to collect. In addition, to build large dataset by downloading the images from search engine is very difficult and most financially challenging. One way to get around a lack of large face datasets is to augment face datasets by synthesizing new possible views of the face images

they contain. Thereby, an existing face set is expanded to many times its size by introducing supplementary intra-subject appearance variations such as pose variations.
The remainder of the paper is as follows. In Section 2 we talks about the related works, Section 3 introduce our new proposed method (Pubface). In Section 4, we present the experiments and discuss our method. Section 5 concludes the paper.

2. Related work

The problem with small datasets is that deep learning models particularly Convolutional Neural Networks (CNNs) trained with them do not generalize well data from the validation and test set. Hence, these models suffer from the problem of overfitting. Therefore, to combat overfitting, many techniques have been proposed for example, dropout [1] and batch normalization [2]. Dropout works by probabilistically removing a neuron from designated layers during training or by dropping certain connection. Batch normalization method, allows us to train the normalization weights and normalizes layers. Batch normalization can be applied to any layer within the net and hence is very effective. Another popular method is data augmentation, where we use the data you currently have and manipulating them to produce more altered versions of the same image, to increase the variety of data seen during training. Data augmentation has shown effective in image classification [3]. Various smart data augmentation methods have been proposed include geometric augmentations such as oversampling [4], rotating [5], mirroring [6] and photometric transformations [7]. In addition to the typical transformations, face image was augmented with Hairstyles synthesis [8], Glasses synthesis [9], Illuminations synthesis [10]. Another framework named Hyper-class augmented and regularized deep learning [11]. This exposes the CNNs models to additional variations without the cost of collecting and annotating more data. This can have the effect of reducing overfitting and improving the model's ability to generalize.

3. Pubface

3.1. Dataset Collection

We have proposed a novel semi-automatic approach to build very large training datasets of synthetic images by compositing real face images in a small dataset with minimal human intervention. Specifically, we present a multi-stage method proposed to build a large synthetic face dataset with over five million faces labelled for identity by employing a smart synthesis augmented approach based on rendering pipeline to increase the pose and lighting variability. The different steps of the proposed approach are summarized below in Table 1; next we provide details about each step.

Step 1: Finding names of celebrities.
Step 2: Finding popular celebrities.
Step 3: Remove overlap with LFW benchmark.
Step 4: Collect representative images for each celebrity
Step 5: Face images were detected using a robust face detector.
Step 6: Remove the copies images of the same image.
Step 7: Annotation process.
Step 8: Synthesizing faces.

Figure 1. Main stages of the dataset building process

- The first step in building the face dataset is to get an initial list of public figures and popular celebrity's names i.e. the people with high representation on search engines, such as actors, football players and politicians to avoid any privacy issue in downloading their pictures. This is done by using the Internet Movie Data Base (IMDB) celebrity list, which contains the most of public figures ranked by popularity.

IOP Conf. Series: Materials Science and Engineering **353** (2018) 012022 doi:10.1088/1757-899X/353/1/012022

- The second step is the application of the filtering process to eliminate any overlap with the standard LFW [12] benchmark by removed any celebrity name appearing in the LFW face dataset in order to make it possible to train on the new dataset and still evaluate fairly on LFW benchmark, resulting in a lists of 4000 identities.
- The third step of the process consist in the collect a set of representative images for each celebrity of the candidate list from sources like Wikimedia Commons, and search engines like Yandex, Google Image Search and Facebook Graph Search.
- The four step of the process consist the running the Viola-Jones face detector [13] on all images collected in the previous stage to obtain the location of the face in each given image and then cropped and rescaled to a standard coordinate system. In order to align the faces we apply the face alignment method [14] to rotate them accordingly to an up-frontal position.
- The five step is to remove the duplicate images of the same image being found at two different Internet locations. This is done by computing one of the descriptors presented on [15] [16] [17] [18] for all image per each identity.
- The sex step in building the face dataset is the manual annotation process of all face images obtained in the previous stage.
- Finally, in order to augment the size of our dataset, by produce multiple views of face images with more intra-subject appearances variations such as pose variations. We consider an extension of the above-proposed method to produce frontal views to minimize variability for better alignment (Test phase) while we do this to augment variability for better capture intra-class appearance variations (Training phase). Therefore, to produce multiple views of each face image I, we begin by applying the facial landmark detector proposed in [19]. Given these detected facial key points we estimate the 6 DOF for the face in the given image I using correspondences between the facial key points $p_i \epsilon IR^2$ and points $P_i \epsilon IR^3$, labeled on a 3D generic face model S. Here, i indexes specific facial landmarks in I and the 3D shape S. we use PnP [20] to estimate extrinsic camera parameters, which gives us a perspective camera model mapping the generic 3D shape S on the image I. As well as the we make new suitable rotation matrices R for novel multiple views by sampling different yaw angles

$$p_i \sim MP_i \tag{1}$$

$$M = A\,[R\ t] \tag{2}$$

3.2. Network architecture

To perform face recognition, we considered the configuration D of the CNN architecture proposed in [21] in our investigation, which is referred as VGG16 trained on ImageNet [4] for image classification as well as for face recognition [22]. It comprise 11 blocks. The eight first blocks are said to be convolutional. All the convolution layers are followed a rectified unit layers and max pooling. The last three blocks are said to be fully connected layers. The output layer is adapted to 4000 output neurons corresponding to the number of classes of the corresponding Input layer. The number of layers in between the Input layer and the final dense layer denotes the depth of the DL model. To benefit from the representation power of the CNN, fine-tuning of this pre-trained model with our new dataset is a necessary step. We then deploy this model on the Raspberry Pi 3, model B (hardware with limited computational performance).

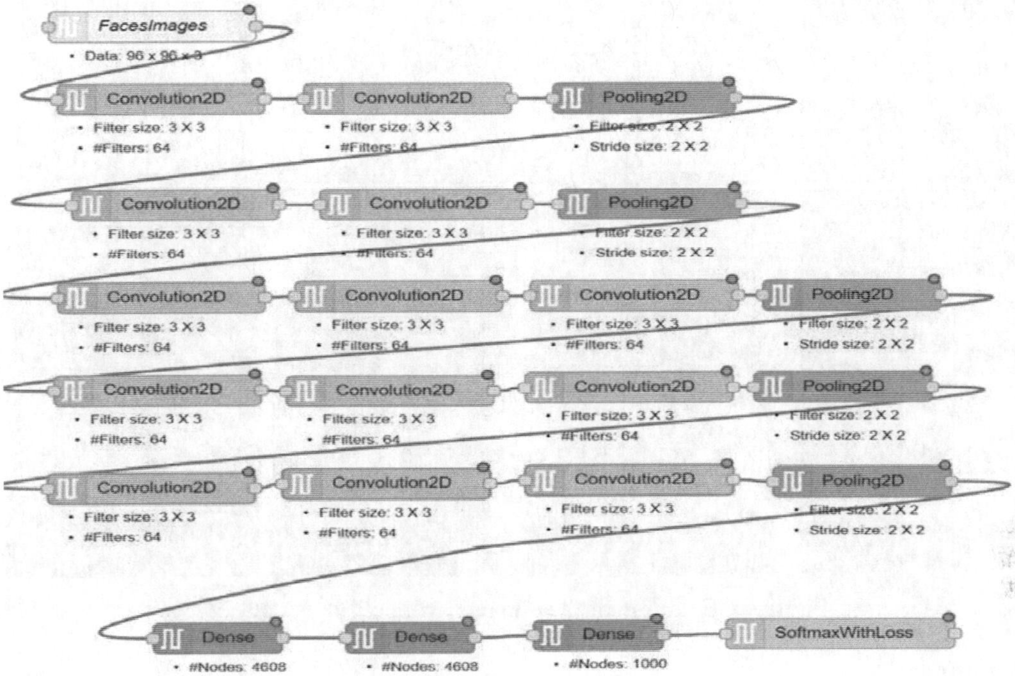

Figure 2. VGG16 architecture

4. Experiments and results

We call the generic dataset building as Puball-dataset, which all the images have the size of 96x96x3 pixels. Table 1 gives some statistical information on the larger face datasets public and private. The whole dataset is split as 60% for training, 20% validation, and 20% for testing.

Table 1. Dataset comparison

Dataset	Identities	Images
Facebook [23]	4,030	4.4M
Google [24]	8M	200M
MegaFace [25]	690,572	1.02M
LFW [12]	5,749	13,233
CelebFaces [26]	10,177	202,599
Puball-dataset (ours)	4000	5M

The entire CNN model was implemented in python using the deep-learning framework TensorFlow. The use of the pre-processing and alignignement procedures prior to classification increased the recognition rate from 90% to 98%. As benchmark we compare the proposed approach to OpenCV classifier was achieved the highest performance known as Fisherface.

Table 2. Benchmark Accuracy rate and Classification time for the proposed model and Fisherface

Method	Accuracy Rate (%)	Classification time (ms)
Fisherface	96	420
Our proposed model (ours)	98	90

Figure 3. Example of face images from our new dataset

5. Conclusion

In conclusion, our deep-learning based approach called Pubface lead to impressive results on hardware with limited resources with respect to both speed and accuracy compared to traditional OpenCV classifier known as fisherface which was achieved the highest performance. Pubface can recognize faces with 98% accuracy in real time, which is pretty much as good as humans can do. It's pretty amazing technology.

References

[1] G. E. HINTON, N. SRIVASTAVA, A. KIRZHEVSKY, I. SUTSKEVER and R. R. Salakhutdinov, "Improving neural networks by preventing co-daptation of feature detectors," arXiv preprint arXiv:1207.0580, 2012.

[2] Y. MA and D. KLABJAN, "Convergence analysis of batch normalization for deep neural nets," CoRR, abs/1705.08011, 2017.

[3] S. C. WONG, A. GATT, V. STAMATESCU and M. D. MCDONNEL, "Understanding data augmentation for classification: when to warp?," CoRR, abs/1609.08764, 2016.

[4] A. KRIZHEVSKY, I. SUTSKEVER and G. E. HINTON, "Imagenet classification with deep convolutional neural networks," Neural Inform. Process. Syst., p. 1097–1105, 2012.

[5] S. XIE and Z. TU, "Holistically-nested edge detection," Proc. Int. Conf. Comput. Vision, 2015.

[6] H. YANG and I. PATRAS, "Mirror, mirror on the wall, tell me, is the error small?," Proc. Conf. Comput. Vision Pattern Recognition, 2015.

[7] K. HE, X. ZHANG, S. REN and J. SUN, "Spatial pyramid pooling in deep convolutional networks for visual recognition," arXiv:1406.4729 [cs.CV], 2015.

[8] L. Luoqi , X. Hui , X. Junliang , L. Si , Z. Xi and S. Yan, "Wow! you are so beautiful today!," Proceedings of the 21st ACM international conference on Multimedia, pp. 3-12, 2013.

[9] W. Yandong , L. Weiyang , Y. Meng , F. Yuli , X. Youjun and H. Rui , "Structured occlusion coding for robust face recognition," Neurocomputing, p. 11–24, 2016.

[10] J. Dalong, H. Yuxiao, Y. Shuicheng, Z. Lei, Z. Hongjiang and G. Wen, "Efficient 3D

reconstruction for face recognition," Pattern Recognition , vol. 38, no. 6, pp. 787-798, 2005.

[11] S. XIE, T. YANG, X. WANG and Y. LIN, "Hyper-class augmented and regularized deep learning for fine-grained image classification," Proc. Conf. Comput. Vision Pattern Recognition, p. 2645–2654, 2015.

[12] G. B. Huang, M. Ramesh, T. Berg and E. Learned-Miller, "Labeled faces in the wild: A database for studying face recognition in unconstrained environments," University of Massachusetts, Amherst, Technical Report, pp. 07-49, 2007.

[13] H. Ouanan, M. Ouanan and B. Aksasse, "Implementation and optimization of face detection frameworkbased on OpenCV library on mobile platforms using Davinci's technology," International Journal of Imaging and Robotics, vol. 16, no. 2, pp. 118-131, 2016.

[14] H. Ouanan, M. Ouanan and B. Aksasse, "Novel approach to pose invariant face recognition," Procedia Computer Science, 2017.

[15] H. Ouanan, M. Ouanan and B. Aksasse, "Myface: Unconstrained Face Recognition," Lecture Notes in Networks and Systems, 2017.

[16] H. Ouanan, M. Ouanan and B. Aksasse, "Gabor-HOG Features based Face Recognition Scheme," TELKOMNIKA Indonesian Journal of Electrical Engineering, vol. 15, no. 2, pp. 331-335, 2015.

[17] H. OUANAN, M. OUANAN and B. Aksasse, "Gabor-Zernike Features based Face Recognition Scheme," International Journal of Imaging and Robotics, vol. 16, no. 2, pp. 118-131, 2016.

[18] H. OUANAN, M. OUANAN and B. AKSASSE, "Face Recognition by Neural Networks based on Gabor filters and Random Projection," International Journal of Mathematics and Computation, vol. 26, no. 2, pp. 30-42, 2015.

[19] H. Ouanan, M. Ouanan and B. Aksasse, "Facial landmark localization: Past, present and future," 4th IEEE International Colloquium on Information Science and Technology (CIST), 2016.

[20] T. Taketomi, K. Okada, G. Yamamoto, J. Miyazaki and H. Kato, "Camera pose estimation under dynamic intrinsic parameter change for augmented reality," Computers & Graphics, vol. 44, pp. 11-19, 2014.

[21] K. Simonyan and A. Zisserman, "Very deep convolutional networks for large-scale image recognition," arXiv preprint arXiv:1409.1556, 2014.

[22] H. Ouanan, M. Ouanan and B. Aksasse, "Face Recognition Using Deep Features," Lecture Notes in Networks and Systems , 2017.

[23] T. Yaniv , Y. Ming , R. Marc'Aurelio and W. Lior , "DeepFace: Closing the Gap to Human-Level Performance in Face Verification," Proc. IEEE Conf. Comput. Vis. Pattern Recognit., pp. 1701-1708, 2014.

[24] F. Schroff, D. Kalenichenko and J. Philbin, "Facenet: A unified embedding for face recognition and clustering," Proceedings of the IEEE Conference on Computer Vision and Pattern Recognition, 2015.

[25] I. Kemelmacher-Shlizerman, S. M. Seitz, D. Miller and E. Brossard, "The megaface benchmark: 1 million faces for recognition at scale," Proceedings of the IEEE Conference on Computer Vision and Pattern Recognition, pp. 4873-4882, 2016.

[26] Y. Sun, X. Wang and X. Tang, "Deep learning face representation from predicting 10,000 classes," Proceedings of the IEEE Conference on Computer Vision and Pattern Recognition, 2014.

[27] G. E. Hinton, N. Srivastava, A. Krizhevsky, I. Sutskever and R. R. Salakhutdinov , "Improving neural networks by preventing co-adaptation of feature detectors," arXiv preprint arXiv, pp. 01-18, 2012.

IOP Conf. Series: Materials Science and Engineering **353** (2018) 012023 doi:10.1088/1757-899X/353/1/012023

Modelling and simulation of "Free Cooling" process applied to building construction

A OUSEGUI[1*] and M ASBIK[2]

[1]Département de Physique, Faculté des Sciences, Université Moulay Ismail, BP 11201 Zitoune, Meknès, Maroc
[2]Equipe de Recherche en Thermique et Energie (ERTE), UM5 ENSET, BP 6207, Avenue des FAR, 10100, Rabat, Maroc

Email: *ousegui@gmail.com

Abstract. Thermal energy storage systems (TES), using phase change material (PCM) in building walls, consists a hot topic within the research community currently. In the present work, a numerical model is developed to simulate free cooling of air-PCM heat exchanger in both charging and discharging steps. The studied case is taken from experimental work. The domain consists in two parallel plates made of Paraffin as PCM, separate by a gap where air circulates. The flow and temperature can be adjusted. The goal is to calculate the temperature of the air at the outlet, in order to analyse the performance of the device. A good agreement was founded between experimental and numerical results. The analysis of the influence of the flow rate on the efficiency of the process confirms a previous works, that the heating flow rate should be higher than cooling one.

1. Introduction

One of the alternatives to decrease the energy demands is the use of thermal energy storage (TES) combined with phase change materials (PCM). In the specific conditions, PCM can absorb heat, store it and release it; improving the gap between energy supply and energy consumption [1]. This type of technique called the latent heat thermal energy storage (LHTES) is an attractive way and has taken much attention for heating and cooling purposes in buildings. Many researches and studies based on numerical simulations or experimental work were focused on the change of indoor air temperature due to installation of PCMs [2]. However, many technical tools and design improvement are required for its commercialization in the building sector. Both experimental and numerical researches have been conducted to unravel parameters influencing this kind of process. Thambidurai et al [3] pointed out that the PCM temperature range, geometry of the PCM container, inlet air temperature, encapsulation thickness and the air flow rate are the main parameters affecting the free cooling performance. The control of this last factor ensures full solidification of the PCM during the charging stage or complete melting during discharging. Arkar et al ([4]-[5]) recommended that air flow rate during the charging process should be three to four times higher than air flow rates during the discharging process. Zalba et al [6] and Saman et al [7] observed however that a higher air flow rate boosts the heat transfer rate and leads to a shortened time phase change in both steps. However while charging, if the air temperature is not less than the subcooling temperature of the PCM, a higher air flow rate is not beneficial according to Waqas and Kumar [8].

Cold outdoor air circulates in the building via free or forced ventilation. The first type is less energy demanding, however, it is very climate dependent (it is not efficient when the wind speed is weak). The second type is more efficient because it ensures a constant ventilation of air conditioning; even if the forced ventilation needs an additional energy for the fan.
In the current work we present a model to simulate the free cooling process, during the both steps. The model is validated via experimental works existing in the literature. Followed by a parametric study concerning the influence of inlet parameters (Temperature and air flow), also, the ambient conditions, on the performance of the process.

2. Studied case and assumptions

The reference case for the validation and the parametric study will be based on the experimental work done by [9], (see Figure 1).

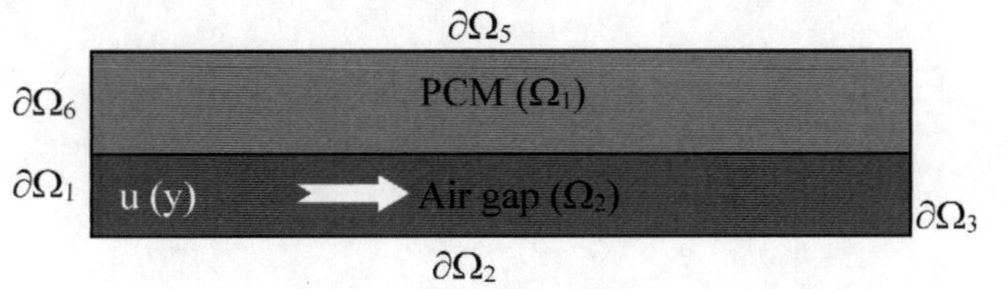

Figure 1 Considered domain

The PCM slab is composed of 6 ENERGAIN®, with thermo-physical properties as summarized in **Erreur ! Source du renvoi introuvable.** [9].

Table 1. Thermo-physical properties of PCM slab

Property	Value
Density	850 (kg/m^3)
Thermal conductivity	0.2 (W/m/K)
Apparent heat capacity	see figure 2
Melting start temperature	13.6°C
Freezing start temperature	23.5°C
Melting peak temperature	22.2°C
Freezing peak temperature	17.8°C

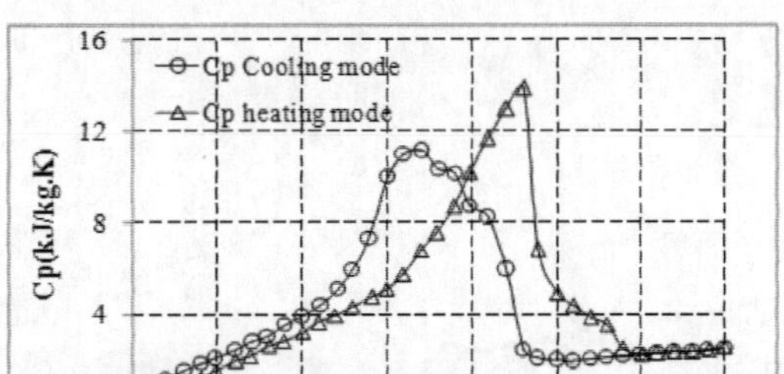

Figure 2. Apparent heat capacity of the PCM [10]

Concerning air, and due to temperature range work we consider constant thermo-physical properties reported at the reference temperature 20°C.
For the velocity field, we suppose that the flow is the fully developed Poiseuil regime:

$$u(y) = \frac{3}{2}\frac{\bar{u}}{H^2}\left(H^2 - y^2\right) \tag{1}$$

where $\bar{u}$ is the mean velocity, and H is the height of the air gap. In addition, because of the small height of the PCM sample, no free convection phenomenon in the liquid state of PCM is considered.
The governing equations of conjugate heat transfer with respect to assumptions considered in are presented in the system below:

$$\rho_s c_{app}(T)\frac{\partial T}{\partial t} = k_s \nabla^2 T \text{ in}(\Omega_1) \tag{2}$$

$$\rho_f c_{pf}\frac{\partial T}{\partial t} + u(y)\cdot\nabla T = k_f \nabla^2 T \text{ in } (\Omega_2) \tag{3}$$

With initial conditions:

$$T_{t=0}=T_0 \tag{4}$$

And boundary conditions

$$\nabla T.n = 0 \text{ at } \partial\Omega_i \; (i=2,6) \tag{5}$$

$$T=T_{in} \text{ at } \partial\Omega_l \tag{6}$$

3. Numerical Method

Here, we used Comsol software (5.0) based on finite element method to solve the system of equations. The structured mesh (quadrilateral elements) technique was adopted, with boundary layers inflation at the air-PCM interface. The mesh consisted of 10500 elements. The time step was automatically performed by the software. At each time step, the system of equations was solved with the linear system solver (PARDISO); with fully coupled option between heat conduction and convections transfers. The apparent specific heat capacity method was adopted in heat transfer equation of PCM to take in the account the phase change phenomenon [11]. Results have been stored every 15 min. The model was run on a 2.7 GHz processor and a 4 Go RAMPC machine.

4. Validation

The validation of the experimental work was focused on the air temperature at the exit of the heat exchanger, for charging and discharging steps; for two flows air 240 and 330 m^3/h.

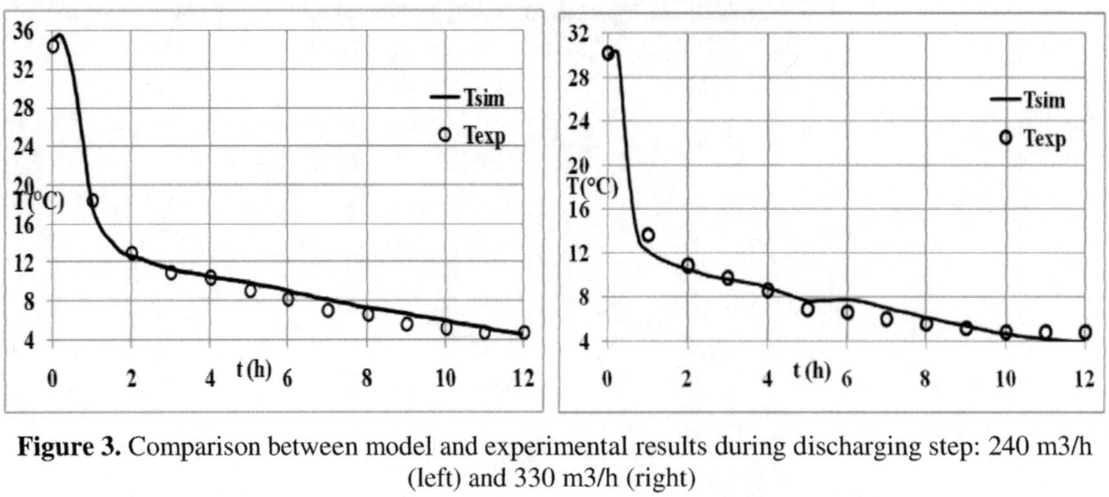

Figure 3. Comparison between model and experimental results during discharging step: 240 m3/h (left) and 330 m3/h (right)

IOP Conf. Series: Materials Science and Engineering **353** (2018) 012023 doi:10.1088/1757-899X/353/1/012023

As can be deduced from **Erreur ! Source du renvoi introuvable.**, the model provides similar results to the experiments for the two processes. The average relative error doesn't exceed 2%. And the maximum error is about 20% for all simulations.

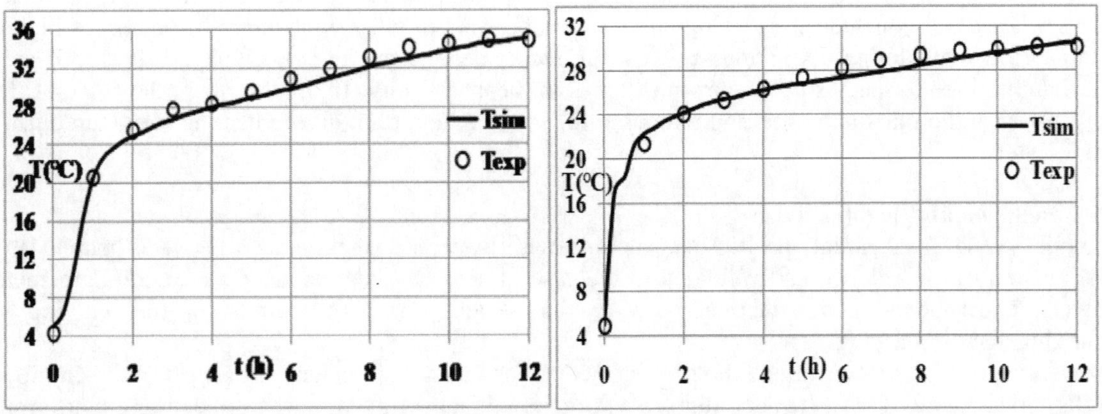

Figure 4. Comparison between model and experimental results during charging step: 240 m3/h (left) and 330 m3/h (right)

5. Performance indicator

During thermal cycling, the key performance indicators are the charging/discharging and overall efficiencies. The charging (discharging) efficiency is the ratio of the stored (recovered) energy in the system (PCM+air) to the net input and pumping energy:

$$\eta_{Charging\ \ or\ discharging} = \frac{E_{stored\ \ or\ recovred}}{E_{net} + E_{pump}} \tag{7}$$

$$E_{stored\ \mathrm{or\ recovred}} = \int_{\Omega} \rho Cp_{app} d\Omega \times (T_{final} - T_{start}) \tag{8}$$

$$E_{net} = \int_0^{t_{process}} \int_{T_{in}}^{T_{out}} \dot{m} C_{p_f} dTdt) \tag{9}$$

$$E_{pump} = \int_0^{t_{end}} \frac{\Delta P}{\rho_a \eta_{fan}} \dot{m} dt \tag{10}$$

Where t_{end} is the duration of charging or discharging process.

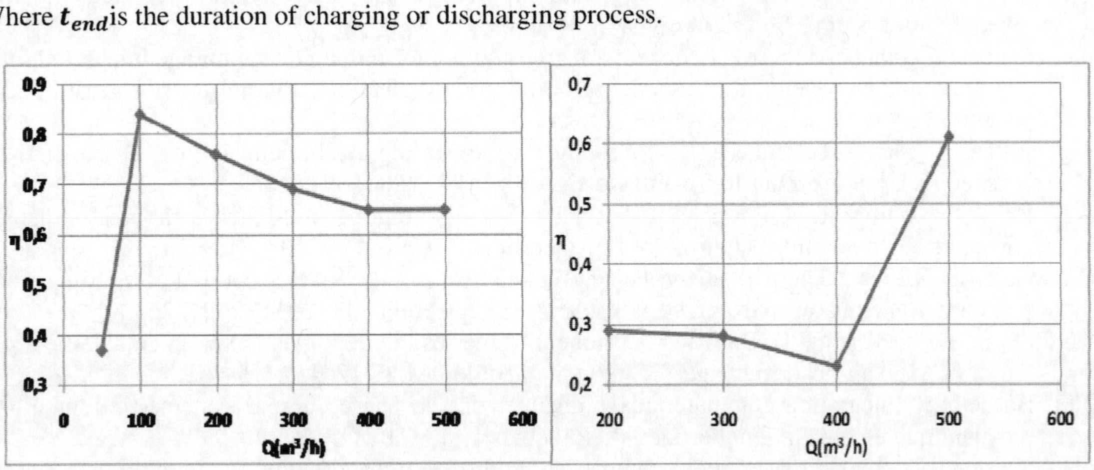

Figure 5. Charging (left) and discharging (right) efficiency for different flow rate

Erreur ! Source du renvoi introuvable. the value of the efficiency of this kind of heat exchanger, in both regimes (charging and discharging). As can be seen, in the discharging process the flow rate reach its optimal value at the flow rate 100 m^3/h. However in the charging step this optimal will be at the flow rate more than 500 m3/h. the same tendency was observed by [4] and [5] in here experimental works where they conclude that during charging process air flow rates of three to four times higher than air flow rates during discharging process are recommended for free cooling applications. This is due mainly to the hysteresis phenomenon of apparent heart capacity (figure 2), where the slope of the Cp_{app} to reach the maximum (melting temperature) is more important during heating step than during cooling step.

6. Conclusion and perspectives

A model based on apparent specific heat method with hysteresis phenomenon between heating and cooling step was developed and validated in the case of air PCM heat exchanger. A good agreement between experimental and numerical value was observed, for both regimes of free cooling ie "charging and discharging".

The parametrical study concerning the flow rate confirms the global tendency observed in the previous works. To simulate the real free cooling process mainly the charging step where the air circulate from the ambient room to heat exchanger (and not from the outdoor), a zonal model simulating the ambient temperature room should be developed and linked to the current model. On the other hand an adequate PCM should be chosen to be applied to the climate of each Moroccan region. Finally the study shows that many parameters can influence the efficiency of the heat exchanger so we believe that an inverse method is essential to optimize the efficiency of the PCM heat exchanger. And finally extending the current model of the free cooling to the "free heating" to e applied during winter season

References

[1] Farah Souayfane, Farouk Fardoun, Pascal-Henry Biwole, Phase Change Materials (PCM) for cooling applications in buildings: A review, Energy and Buildings

[2] E. Osterman, V.V. Tyagi, V. Butala, N.A. Rahim, U. Stritih. Review of PCM based cooling technologies for buildings. Energy and Buildings 49 (2012) 37–49

[3] M. Thambidurai, K. Panchabikesan, K. N. Mohan, V. Ramalingam, "Review on phase change material based free cooling of buildings-The way toward sustainability", Journal of Energy Storage. 4 (2015) 74-88.

[4] C. Arkar, S. Medved, "Free cooling of a building using PCM heat storage integrated into the ventilation system", Solar Energy. 81(10) (2007) 78–87.

[5]C. Arkar, B. Vidrih, S. Medved, "Efficiency of free cooling using latent heat storage integrated into the ventilation system of a low energy building", International Journal of Refrigeration. 30 (2007) 134–43.

[6] B. Zalba, J. Marin, L. Cabeza, H. Mehling, "Free-cooling of buildings with phase change materials", International Journal of Refrigeration. 27 (2004) 39–49

[7] B. Zalba, J. Marin, L. Cabeza, H. Mehling, "Free-cooling of buildings with phase change materials", International Journal of Refrigeration. 27 (2004) 39–49

[8] A. Waqas, S. Kumar, "Thermal performance of latent heat storage for free cooling of buildings in a dry and hot climate: an experimental study", Energy Build. 43 (2011) 2621–2630

[9] J. P. A. Lopez, F. Kuznik, D, Baillis, J, Virgone, "Numerical modeling and experimental validation of a PCM to air heat exchanger", Energy and Buildings. 64 (2013) 415-422.

[10] J. Borderon, "Intégration des matériaux à changement de phase comme système de régulation dynamique en rénovation thermique" [PhD thesis]. ENTPE: Lyon; (2012)

[11] A. Ousegui, A. Le Bail, M. Havet, "Numerical and experimental study of a natural convection thawing process", AIChE J, 52 (2006) 4240–4247

IOP Conf. Series: Materials Science and Engineering **353** (2018) 012024 doi:10.1088/1757-899X/353/1/012024

Implementation of Maximum Power Point Tracking (MPPT) Solar Charge Controller using Arduino

B Abdelilah[1*], A Mouna[1,2], N KouiderM'Sirdi[2], A El Hossain[1]

[1] Laboratoire Signaux ,Système et composants(LSSC). FST. Fès. Maroc.
[2] Laboratoire des Sciences de l'information et des systčmes(LSIS). AMU France.

Email: byou_2009@yahoo.fr

Abstract— the platform Arduino with a number of sensors standard can be used as components of an electronic system for acquiring measures and controls. This paper presents the design of a low-cost and effective solar charge controller. This system includes several elements such as the solar panel converter DC/DC, battery, circuit MPPT using Microcontroller, sensors, and the MPPT algorithm. The MPPT (Maximum Power Point Tracker) algorithm has been implemented using an Arduino Nano with the preferred program. The voltage and current of the Panel are taken where the program implemented will work and using this algorithm that MPP will be reached. This paper provides details on the solar charge control device at the maximum power point. The results include the change of the duty cycle with the change in load and thus mean the variation of the buck converter output voltage and current controlled by the MPPT algorithm.

Keywords — Solar Energy, MPPT, Charge controller, Arduino Nano, Battery.

1. Introduction

Global demand for energy is rapidly evolving and natural energy resources such as uranium, petroleum, and gas decreased due to a great diffusion and development of the industry in recent years. The increase in energy costs and environmental constraints are pushing for the development of technological solutions allowing better control of the resources and the exploitation of the renewable energies in specific photovoltaic energy. Photovoltaic energy is a clean and renewable energy resource. Moreover, solar panels are a silent energy producer because there is absolutely no noise when converting sunlight into electricity [1].

In order to exploit solar energy to power the DC loads and to store electricity, a solar charge controller is needed to monitor the State of charge of the batteries and protect them from overcharging and full discharge (deep discharge). This monitoring and this permanent protection help to extend significantly the performance and life of the batteries. This controller is used in many areas such as systems not connected to the electric network, ensure the autonomy of an embedded system, monitor solar installations...

IOP Conf. Series: Materials Science and Engineering **353** (2018) 012024 doi:10.1088/1757-899X/353/1/012024

This paper includes an MPPT circuit in order to extract the maximum power from the solar panel using the algorithm P & O (perturbation & Observation) and the DC /DC Converter.

2. Proposed system

The block diagram shown in figure 1 represents the complete system. It contains a solar panel, DC-DC buck converter, Arduino Nano, battery, and Loads. The solar panel is used to generate the current and voltage from which the voltage will be converted by a DC-DC converter. After calculating the power from the data transmitted by the current and voltage sensors, we use perturb and observe algorithm to reach the maximum power point.

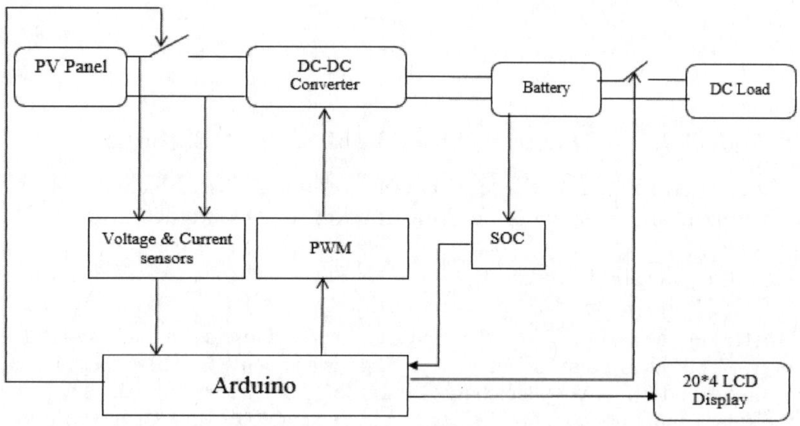

Figure 1. Block diagram of the whole system

2.1 DC-DC Buck Converter

The buck converter produces a lower average output voltage than DC input voltage for charging the battery and supplying the DC loads. The DC-DC buck converter circuit is shown in Figure 2.

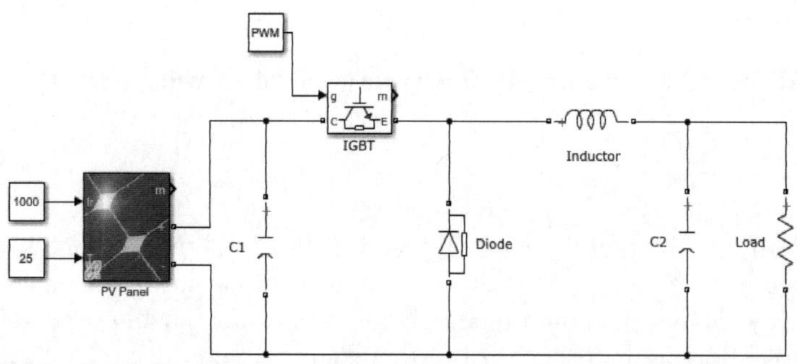

Figure 2. DC-DC Buck converter

After designing the buck converter, we need to find the values of the inductor and the capacitor. It is most critical part as the efficiency of the converter depends on these values. For a solar panel of 60W and a 12V battery, the value of the inductance L = 33uH and the capacitance value C = 220uF, [2].

2.2 MPPT Algorithm

Maximum Power Point Tracking (MPPT) algorithms are needed to maximize the power output at any moment. Many MPPT techniques have been proposed in the literature. In this controller, we chose worked with the method perturb and observe due to its easy implementation. It uses the P-V

IOP Conf. Series: Materials Science and Engineering **353** (2018) 012024 doi:10.1088/1757-899X/353/1/012024

characteristics P = f (V) of the PV module shown in figure 3. Note that the point of maximum power P (n) =V (n) I (n) is obtained when the condition $\frac{dP}{dV}$ = 0 is accomplished, regardless of the solar irradiance,[3]. In this work polycrystalline Silicon panel is used. Electrical characteristics of the PV Panel (Values at STC (AM1.5, 1000W/m², 25°C))

- Maximum Power (Pmax): 60W
- Open-circuit voltage (Voc): 26.4V
- Short-circuit Current (Isc): 3.00A
- Maximum Power Voltage (Vmp):22.0V
- Maximum Power Current (Imp):2.73A
- Tolerance (Tol): +3/-3%

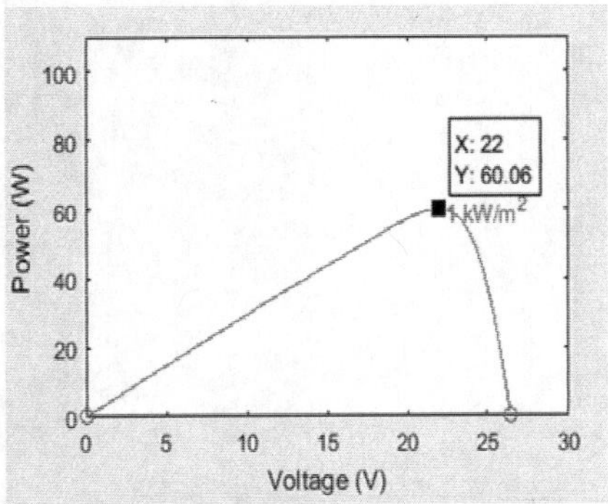

gure 3. P-V characteristic curve Ppv = f (V) of PV modules

The figure 4 shows the Flowchart of the Perturb and Observe method.

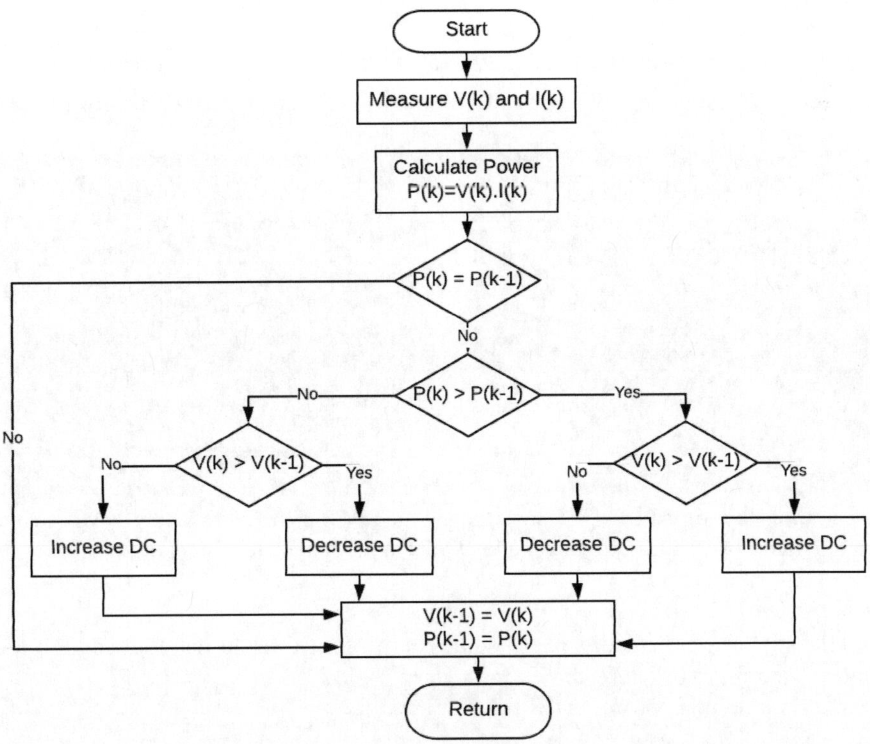

Figure 4. The flowchart of the P&O method

IOP Conf. Series: Materials Science and Engineering **353** (2018) 012024 doi:10.1088/1757-899X/353/1/012024

3. Experimental Results

A prototype was built using various electronic components, this prototype was tested block by block before the realization of the global circuit. The DC-DC converter is the most critical part, the experimental results are presented in figure 6. The LCD screen was used to display the results obtained. The observations ahead show the results of the hardware implementation.

Figure 5. Hardware Prototype

3.1 Buck Converter

In order to verify the design efficiency of the solar charge controller system, an experimental DC-DC Buck converter was built. The PV array was replaced by a battery of 12V. The PWM signal was generated with Arduino Nano board.

The output voltage can be controlled by the duty cycle with the following expression: $\alpha = \frac{Vout}{Vin}$

Figure 6 shows the PWM signal and the output voltage of the DC-DC converter for the different Duty cycle.

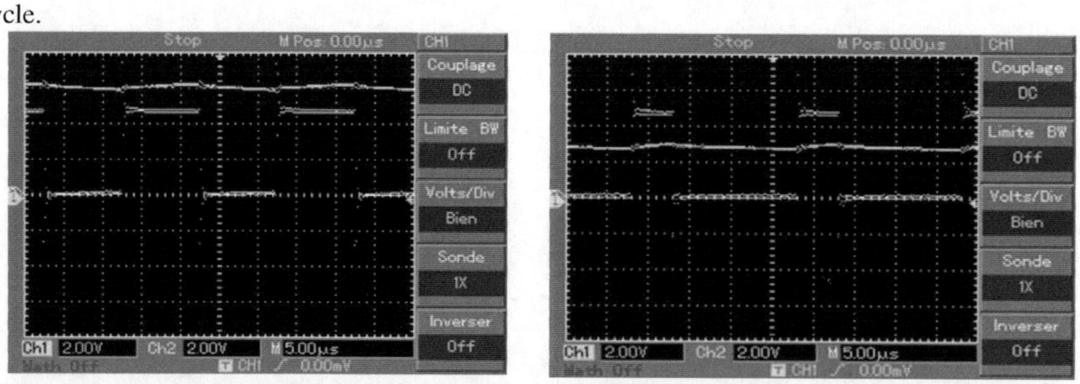

Figure 6. a) Output voltage of the DC-DC converter and the PWM signal with a duty cycle equal to 0.5 and b) Output voltage of the DC-DC converter and the PWM signal with a duty cycle equal to 0.5

3.2 Charge Controller

Table 1, shows the experimental results of the solar charge controller. It has four states on, off, bulk and float, details of the four states are shown in the table below.

IOP Conf. Series: Materials Science and Engineering **353** (2018) 012024 doi:10.1088/1757-899X/353/1/012024

Table 1. Charge controller results

State	conditions required	result	observations
ON	sol_power > Low_sol_power & sol_power < Min_sol_power	Load On & Battery charging On	
OFF	solar watts < Low_SOL_WATTS	Load On & Battery charging Off	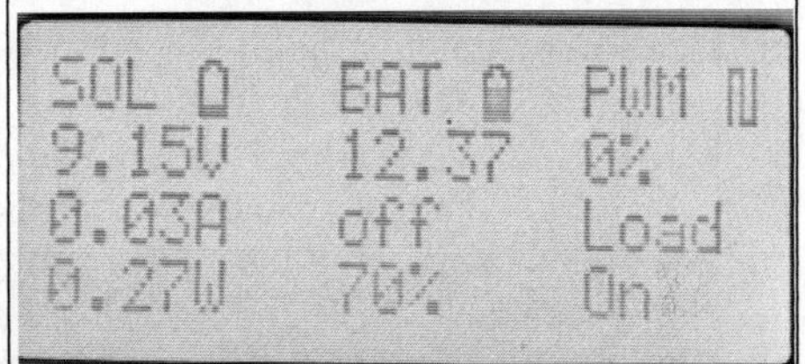
Bulk	solar watts > MIN_SOL_WATTS & bat_volts < BATT_FLOAT	Load On & Battery charging Bulk	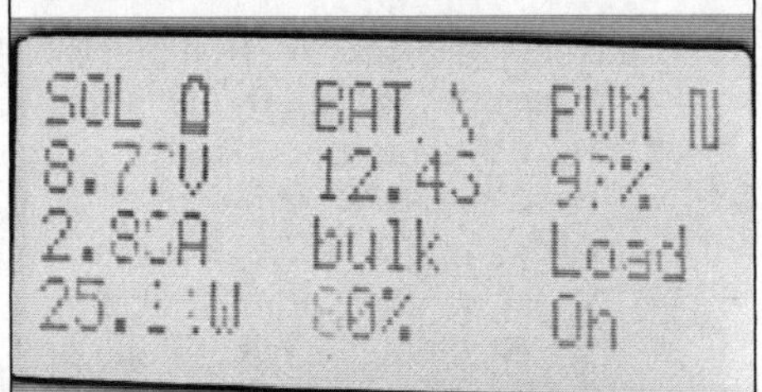
Float	bat_volts > BATT_FLOAT	Load On & Battery charging Float	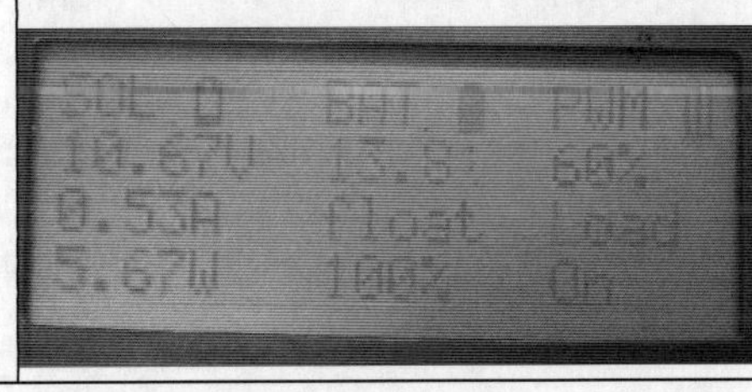

ON State - the solar watts input is between the low and minimum solar power but not low enough to go into the off state. In this state, the Load turns OFF and the Battery charging turns ON.
OFF State - The charger controller goes into this state when there is no more power being generated by the solar panels. In this state, the Battery charging turns OFF and the Load turns ON if the battery voltage is higher than the minimum value.
BULK State – The Solar energy is greater than the minimum value. This is where we do the bulk of the battery charging and when we execute the Peak Power Tracking algorithm. In this state, we try to run the maximum current that the solar panels generate for the battery. The Load turns ON and the Battery charging goes into the Bulk state.
FLOAT State – the battery voltage gets to the maximum Value. In this state, we try and keep the battery voltage at maximum by adjusting the PWM value. The Load is turned ON and the Battery charging goes into the Float state.

4. Conclusion

The realization of this solar charge controller using Arduino was done at low cost but effective for small DC loads, and we can use this controller to implement other algorithms or to make comparisons. This paper contains a battery charging system for laboratory using a direct connection between the photovoltaic solar panel and the battery system. With the help of this charge controller, the authors used solar power effectively and prolonged battery lifetime. In this paper, we have presented the experimental results using the LCD display (panel power, state of charge of the battery, state of the charge controller, duty cycle ....).

Acknowledgments

The authors would like to give very special thanks to Debasish Dutta. As I have learned and used several things from her project.

5. References

[1]. W. Makni, N. Ben hadj, H.Samet, R.Neji, Design simulation and realization of solar battery charge controller using Arduino Uno, a 17th international conference on Sciences and Techniques of Automatic control & computer engineering- STA'2016, Sousse, Tunisia, December 19-21, 2016.

[2]. ON Semiconductor, LC Selection Guide for the DC-DC Synchronous Buck Converter, AND9135/D.

[3]. N K. M'Sirdi, B. Nehme, M. Abarkan, A. Rabhi, The Best MPPT Algorithms by VSAS approach for Renewable Energy Sources (RES), conference paper EFEA 2014, PARIS.

[4]. M. Pathare, V. Shetty, Designing and Implementation of Maximum Power Point Tracking(MPPT) Solar Charge Controller, 2017 International Conference on Nascent Technologies in the Engineering Field (ICNTE-2017).

[5]. N. Barua, A. Dutta, S. Chakma, Implementation of Cost-effective MPPT Solar Photovoltaic System Based on the Comparison between Incremental Conductance and P&O Algorithm, 2016 IEEE International WIE Conference on Electrical and Computer Engineering (WIECON-ECE) 19-21 December 2016, India.

[6]. A. Chandwani, A. Kothari, Design, Simulation and Implementation of Maximum Power Point Tracking (MPPT) for solar based Renewable Systems, 2016 International Conference on Electrical Power and Energy Systems (ICEPES). Dec 14-16, 2016.

[7]. T. Kaur, J. Gambhir, S. Kumar, Arduino Based Solar Powered Battery Charging System for Rural SHS, Power Electronics (IICPE), 7th 2016 India International Conference.

IOP Conf. Series: Materials Science and Engineering **353** (2018) 012025 doi:10.1088/1757-899X/353/1/012025

Energy Management of An Extended Hybrid Renewable Energy System For Isolated Sites Using A Fuzzy Logic Controller

Sanaa Faquir[1,2], Ali Yahyaouy[2], Hamid Tairi[2] and Jalal Sabor[3]

[1] Laboratoire Systèmes et Environnements Durables (LSED), Faculté des Sciences de l'Ingénieur (FSI), Université Privée de Fès (UPF), Fez, Morocco
[2]LIIAN Laboratory, Computer Science Department, College of Science Dhar Mehraz, Université Sidi Mohamed Ben Abdallah, Fez, Morocco
[3]Department of AEEE, Ecole Nationale Supérieure d'Arts et Métiers (ENSAM), Meknes, Morocco

Email: sanaa.faquir@usmba.ac.ma

Abstract. This paper presents the implementation of a fuzzy logic controller to manage the flow of energy in an extended hybrid renewable energy system employed to satisfy the load for a wide isolated site at the city of Essaouira in Morocco. To achieve Efficient energy management, the system is combining two important renewable energies: solar and wind. Lithium Ion batteries were also used as storage devices to store the excess of energy provided by the renewable sources or to supply the system with the required energy when the energy delivered by the input sources is not enough to satisfy the load demand. To manage the energy in the system, a controller based on fuzzy logic was implemented. Real data taken from previous research and meteorological sites was used to test the controller.

1. Introduction

Recently, governments and industries all around the world are progressively trying to find alternatives to overcome the significant increase in the prices of conventional energy sources and to minimize the greenhouse gas emissions (GHC) and the resulting global warming. For these reasons, the implementation and diffusion of renewable energies as well as the establishment of a low-carbon economy have been subject of research in many regional and international initiatives, institutions and financial mechanisms [1]. The combination of solar and wind energy sources along with a storage unit, such as batteries, offers an excellent solution to problems caused by the stochastic nature of these sources, using the strengths of a source to overcome the weakness of the other [2] especially in developing countries where the economic development considers social and environmental issues [3].

The country of Morocco, Among the Northern African countries, is extremely depending on energy imports and fossil fuels. The electricity demand is increasing and therefore, the country will be facing in the future high energy costs if electricity prices continue to increase. To overcome these problems, the country started implementing wind farms and PV plants over big surfaces in different places in Morocco

IOP Conf. Series: Materials Science and Engineering **353** (2018) 012025 doi:10.1088/1757-899X/353/1/012025

as mentioned in [4] [5]. However, there are only few reports on the performance of standalone hybrid systems implemented in small and large isolated sites in the country. This is due to the energy management of these systems that consists on controlling the energy between the input and output sources of energy while satisfying the load demand. Many methods were used before to control the flow of energy in PV/Wind/batteries systems as mentioned in [2], [6], [7]. However, no systems were used to manage the energy in large sites knowing a high rate of poverty and electricity problems.

This paper presents an implementation of a fuzzy logic controller to manage the flow of energy in a PV-wind-battery system for a large isolated site in the city of Essaouira in Morocco composed of 10 residential compounds and four olive oil factories providing a maximum of 850KW of energy per day. The rural electrification is also included in the load demand.

The energy management controller was implemented by an open source Java library called jFuzzyLogic that proposes a fully functional and complete implementation of a fuzzy inference system (FIS), and provides a programming interface (API) and an Eclipse plug-in in order to make it easier to design and test FCL code.

The fuzzy controller was tested on real data taken from the city of Essaouira which is known by its important solar radiation and its average annual wind speed of 8m/s. Results showed that the controller was able to manage the energy in the hybrid system while maintaining the batteries state of charge between 30% and 80%.

2. Fuzzy logic control

Although other intelligent methods such as genetic algorithms and neural networks can perform just as well as fuzzy logic in many applications, fuzzy logic has the advantage that the solution to the problem can be cast in terms that human operators can understand (using linguistic variables to represent vague concepts), so that their experience can be used in the design of the controller. It is also used when the available process model is no precise or when the analytical model is too difficult to evaluate. Thus, it can be adopted for effective energy planning. Fuzzy logic controllers (FLC) were; therefore, used in many applications to manage the energy in hybrid renewable energy systems (HRES) as in [8] and [9]. It was shown that FLCs were more efficient when applied compared to other methods.

3. PV/Wind/Battery stand-alone system for a large isolated site

The system proposed in this paper will be providing the load demand for a large isolated site that includes 10 compounds of 16 villas each and 4 factories in the city of Essaouira. The system also includes the rural lighting. To decide about the number of input and output units required to implement the HRES, the load demand should be defined.

3.1 Load demand acquisition

Given all possible appliances used in a house, the load demand was classified by priority into [6]. Since the load demand by priority varies hourly depending on the need of the electrical appliance by the user at a certain period of the day, the total load demand required was calculated for each hour.

The proposed site includes 10 compounds with an average hourly load demand equals to 26.16KW a minimum equals to 22 KW and a maximum of 36 KW. It includes also four olive oil extraction factories consuming between 80 and 352 KW hourly and with an average hourly load demand equals to 229 KW. As mentioned earlier, the public lighting will be also covered by the proposed hybrid system. The hourly load demand for it is 25KW when the lights are on and this will happen between approximately 7 pm and 6 am. During the other hours, the public lights will not consume any energy.

Taking into account all the elements of the site, the total load demand will then vary between 332.2KW and 738.8 KW with an average rate of 538KW as shown in the figure below:

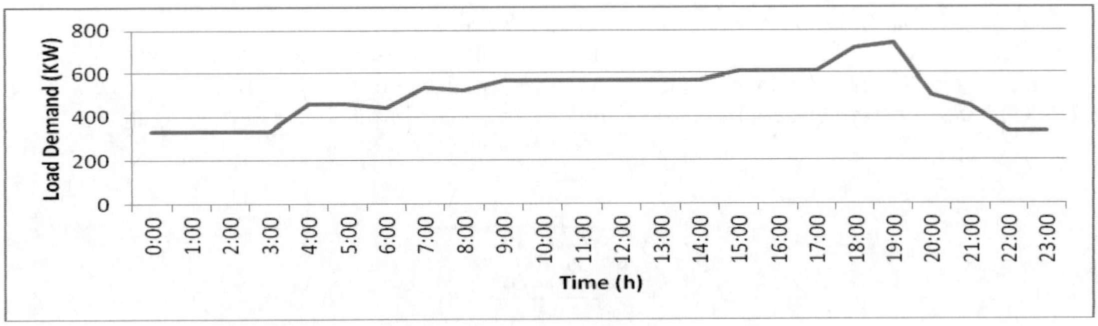

Figure 1. Variation of the load demand for the proposed site per day

3.2 Configuration of the proposed stand-alone PV/Wind/Battery system

The hybrid system proposed is composed of the following units:

- **PV panels***:* 1600 unit implemented in series covering an area of 1km^2 to deliver a maximum power of 100 KWh per day under STC conditions and 18% efficiency.
- **Wind Turbine***:* of type GAMEZA 52 with a maximum power of 850KWh. The performance coefficient (Cp) used in the wind equation was modified to allow the turbine to keep retrieving a maximum power even when the wind speed is too high without causing the turbine to stop [10].
- **Batteries***:* Different storage units were compared in [11] and storage units that were found more convenient to HRES are batteries. The total number of 48V batteries required by the system is 167 Batteries with a maximum power 1600KW (200 Ah each).

4. Energy Management for the HRES proposed using fuzzy logic control

In general, load management means maintaining a balance between the energy produced by the input sources and the electrical load demand while protecting the storage units (batteries) from excessive charge or discharge [6]. The system was tested with load demand statistics (figure 1) and weather data [12] taken in the city of Essaouira in Morocco with full batteries.

4.1 PV and WT Power estimation

PV Power and Wind power were predicted using a fuzzy modeling system to overcome problems related to complex mathematical formulas and imprecise coefficients as described in [13]. Results are shown in figures 2 and 3.

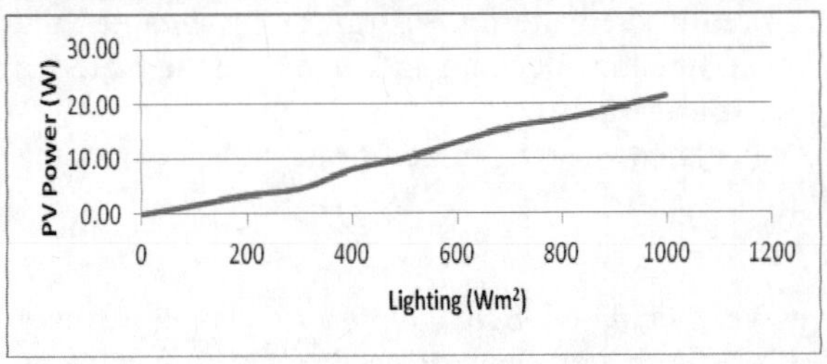

Figure 2. PV power for one panel under temperature=25°C

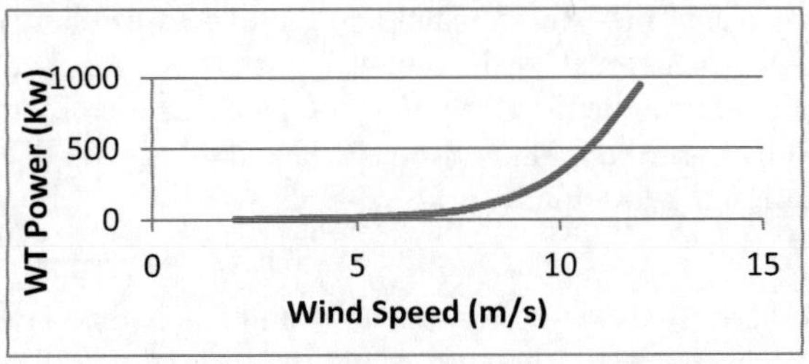

Figure 3. WT power curve

IOP Conf. Series: Materials Science and Engineering **353** (2018) 012025 doi:10.1088/1757-899X/353/1/012025

4.2 The flowchart for the fuzzy controller proposed to the HRES

To ensure the management of the power between the renewable sources and the batteries, satisfy the load demand, and protect batteries from overcharge and over discharge (battery's state of charge (SOC) between 0.3 and 0.8) a fuzzy logic controller was implemented. The flowchart is shown in figure 4.

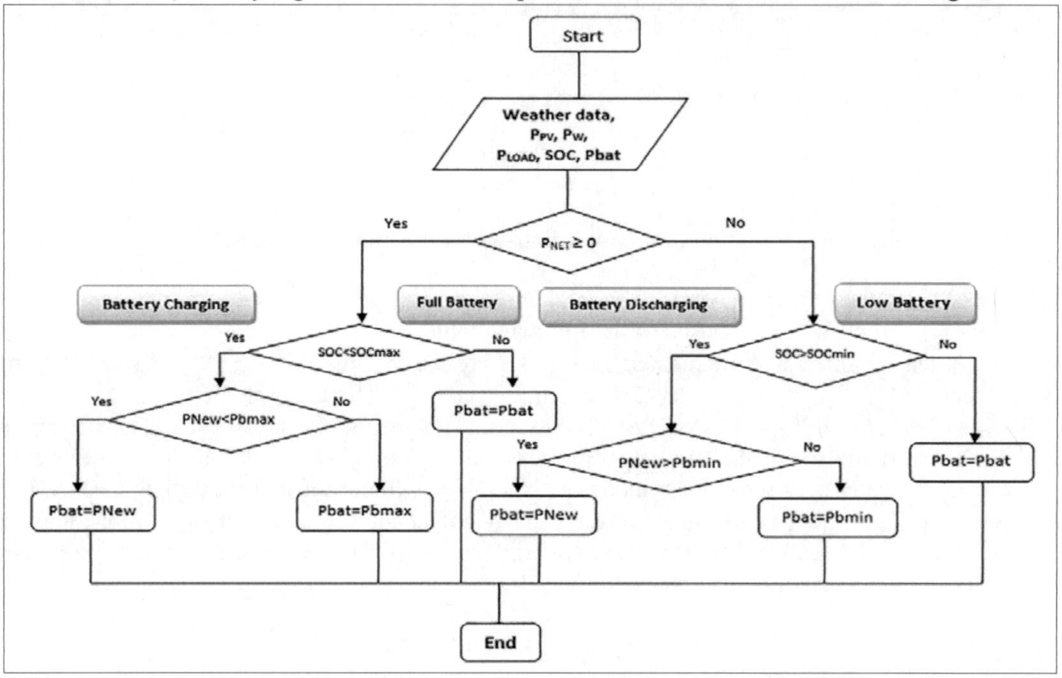

Figure 4. Flowchart representing the energy management system based on type-1 fuzzy control

P_{sys} is the system power calculated from the power generated by the PV panels (P_{pv}) and the power generated by the Wind turbine (Pw) at each period of time (t):

$$P_{sys}(t)=P_{pv}(t)+P_{w}(t) \quad (1)$$

$$P_{Net}(t)= P_{sys}(t)-P_{load}(t) \quad (2)$$

P_{load} represents the load power demand and it will be obtained from the data in figure 1.

The Net power (N_{et}) can be either positive: the power generated by the renewable sources is fully satisfying the load demand and the excess of energy is stored in the batteries (charging), or negative: the power generated by the renewable sources cannot fully satisfy the load demand and therefore, the batteries should provide the remaining energy demanded by the load (discharging).

SOC should remain between 30% and 80%. The SOC is calculated at each period of time t (1 hour) by the following formula:

$$SOC(t) = \frac{P_{bat}(t)}{BC} \quad (3)$$

Where BC is the battery capacity and P_{bat} is the battery power calculated, SOC_{min}=30% (the battery cannot be discharged if it is at its minimum level), SOC_{max}=80% (the battery cannot be charged if it is at its maximum level). The maximum and minimum level of the batteries are then calculated as follows:

$$Pb_{min}=SOC_{min}*BC \quad (4)$$

$$Pb_{max}=SOC_{max}*BC \quad (5)$$

P_{new} shows the battery level that it is supposed to reach after adding the P_{Net} and is calculated at each period of time (t) by the following formula:

$$P_{new}(t)=P_{Net}(t)+P_{bat}(t-1) \quad (6)$$

- **Case of battery charging:**

When $P_{Net}>0$, batteries will be charged if $SOC<SOC_{max}$. SOC_{new} is calculated from P_{new} .
If $SOC_{new}<SOC_{max}$ then P_{Net} will be charged totally and therefore $P_{bat}(t)=P_{new}(t)$.
If $SOC_{new}>SOC_{max}$ then P_{Net} will be charged partially therefore $P_{bat}(t)=Pbmax$

- **Case of battery discharging:**

In case $P_{Net}<0$, batteries will be discharged if $SOC>SOC_{min}$,
If $SOC_{new}>SOC_{min}$ then P_{Net} will be taken totally from batteries, $P_{bat}(t)=P_{new}(t)$.
If $SOC_{new}<SOC_{min}$ then P_{Net} will be taken partially, $P_{bat}(t)=Pbmin$

4.3 Application of the Fuzzy controller

Fuzzy control language (FCL) was incorporated in the Jfuzzylogic package (IEC 61131 part 7) to implement the phases of the FLC (Fuzzification, Inference rules, and deffuzification).

4.3.1. *Fuzzification phase*

- *Input linguistic variable:* $\mathbf{P_{Net}}$, (P (Positive), N (Negative)), **SOC**(Low, Medium, High), $\mathbf{P_{New}}$ (VeryLow, Low, Medium, High, VeryHigh)
- *Output linguistic variable:* **Bat_status** (VeryLow, Low, Medium, High, VeryHigh)

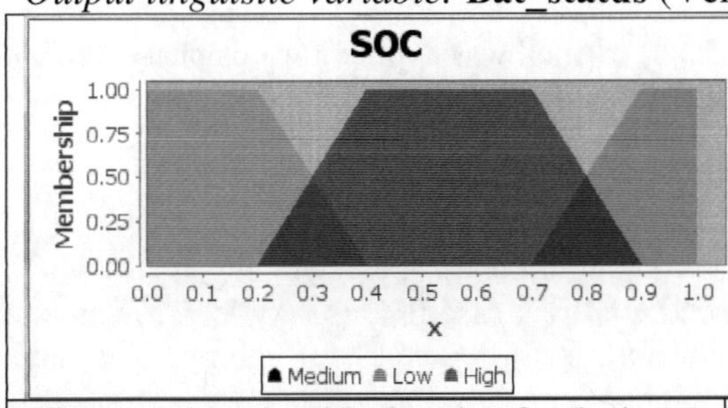

Figure 5. Membership function for the input variable SOC

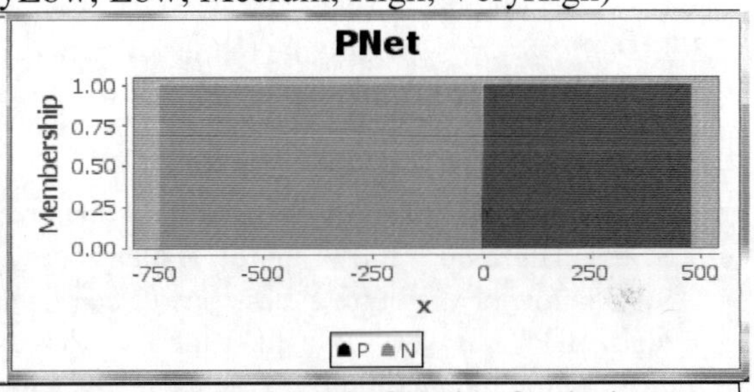

Figure 6. Membership function for the input variable P_{Net}

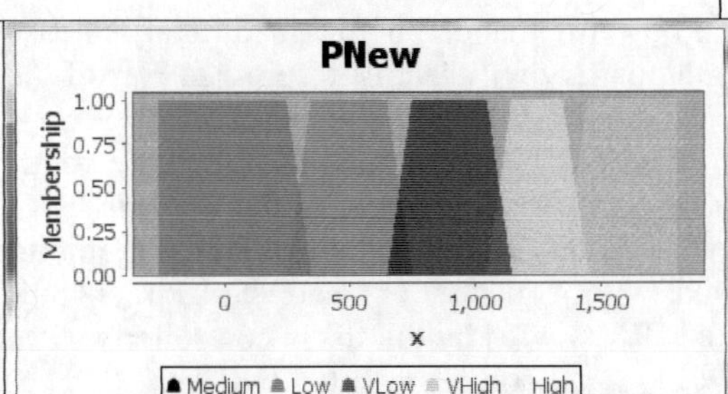

Figure 7. Membership function for the input variable P_{New}

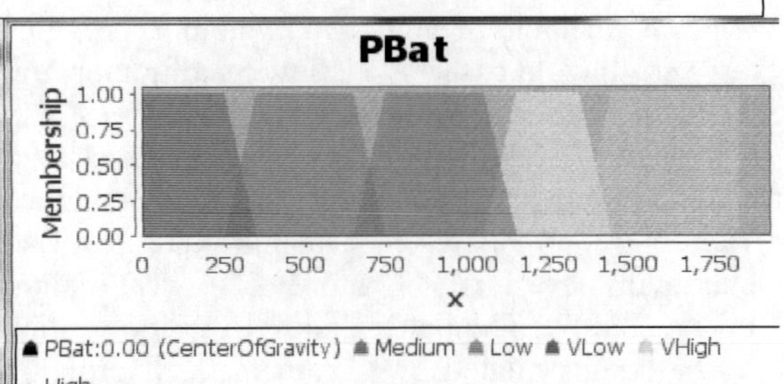

Figure 8. Membership function for the output variable Battery Status

IOP Conf. Series: Materials Science and Engineering **353** (2018) 012025 doi:10.1088/1757-899X/353/1/012025

Figures 5,6,7 and 8 show the membership functions of the input variables SOC, P_{Net}, and P_{New} and the output variable P_{bat}.

4.3.2. *Inference Rules.* The inference rules were interpreted by the De Morgan's Law (min-max for the activation and accumulation methods) to determine the Battery status as shown in Table 1.

Table1. Inference rules of the fuzzy controller

P_{Net}	SOC	P_{New}	P_{bat}	P_{Net}	SOC	P_{New}	P_{bat}
N	Medium	VeryLow	Low	P	Low	Low	Low
N	Medium	Low	Low	P	Low	Medium	Medium
N	Medium	Medium	Medium	P	Low	High	High
N	Medium	High	High	P	Low	VeryHigh	High
N	Medium	VeryHigh	High	P	Medium	Low	Low
N	High	VeryLow	Low	P	Medium	Medium	Medium
N	High	Low	Low	P	Medium	High	High
N	High	Medium	Medium	P	Medium	VeryHigh	High

4.3.3. *Deffuzification phase.* The Center of gravity (COG) method was applied for this phase, the battery level was determined by a crisp value for each period of time t (1h).

5. Results and discussions

In case of P_{Net} is P (Positive) the following rules were derived:

- SOC is High (80% full or more): Battery kept at its maximum level of charge.
- SOC is low or medium, excess power can be stored totally in case P_{New} is very low, low, medium, or high. If P_{New} is very high then the battery cannot absorb all the excess of energy. The battery can absorb until its maximum is reached ($SOC=SOC_{max}=80\%$).

In case of P_{Net} is N (Negative), the following rules were derived:

- SOC Low: battery level at its maximum depth of discharge.
- SOC medium or high, battery able to provide the power demanded by the load totally in case P_{New} is very low. In case P_{New} is low, medium, or high, the battery will be able to provide part of the energy demanded until its minimum is reached ($SOC=SOC_{min}=30\%$).

Given all the rules above, the battery's level fuzzy value can be determined by the fuzzy controller for each case depending on its state of Charge and the energy provided by the renewable sources.
The controller was tested with full batteries of a total capacity of 1600kw. The batteries minimum and maximum levels (Pb_{min} and Pb_{max}) were calculated with equation (4) and (5): Pb_{min}=480KW and P_{max}=1280KW. The battery's level variation estimated ($P_{bat\_estimated}$) by the fuzzy controller was compared to the one calculated by a mathematical formula ($P_{bat\_theoretical}$) and represented by figure 9. The chart is showing a convergence which means that the proposed fuzzy controller gave good estimation to the batteries level for each period of time.

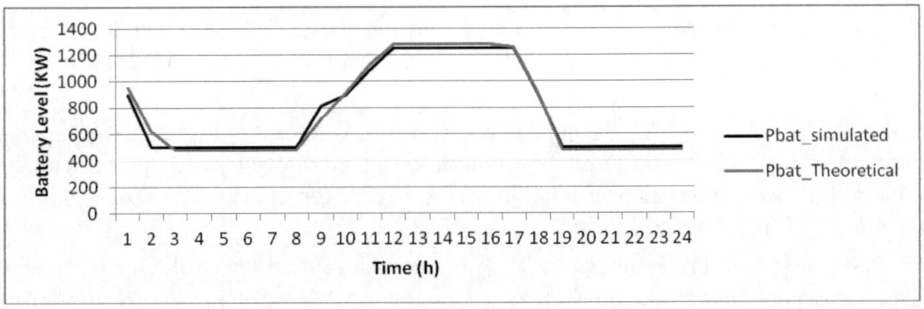

Figure 9. Battery status comparison between the theoretical calculation and the fuzzy estimation

6. Conclusion

This paper presented a stand-alone (off-grid) hybrid system combining solar and wind sources of energy with batteries in one system to satisfy the electrical load demand for a large isolated site in the city of Essaouira in Morocco. A controller based on fuzzy logic was proposed in this paper to manage the flow of energy in the system and determine the status of the battery at each period of time while maintaining the battery's SOC level between 30% and 80%. The algorithm was tested using real data for the region captured for one day in the month of April. Results showed that the fuzzy controller presented an efficient way to control the flow of energy between input sources and storage units while ensuring a continuous power supply for the load demand. The variation of values shown is due to the choice of the fuzzy sets. Since complicated systems may require several iterations to find a set of rules resulting in a stable system, other models and techniques can help in finding an optimal action-selection policy for any given Markov decision process (MDP). These models can be functions such as parametric functions or algorithms such as Q-learning [14].

7. References

[1] K. Fritzsche, D. Zejli and D. Tanzler, 2011, "The relevance of global energy governance for Arab countries: The case of Morocco," *Energy Policy,* vol. 39, no. 8, p. 4497–4506.

[2] M. Seddik, S. Zougga, M. Oukili, T. Ouchbel, A. Aziz, M. L. Elhafyani and F. K. Kadda, 2013, "The Digital Energy Management of a Stand-Alone Hybrid System Photovoltaic-Wind," *International Journal of Engineering and Innovative Technology (IJEIT),* vol. 3, no. 3, pp. 156-163.

[3] B. K. Koua, P. M. E. Koffi, P. Gbaha and T. Siaka, 2015, "Present status and overview of potential of renewable energy in Cote d'Ivoire," *Renewable and Sustainable Energy Reviews,* vol. 41, p. 907–914.

[4] Aderee, 2009, "Energie Eolienne," Agence de Développement des Energies Renouvelables et de l'Efficacité Energétique, Rabat-Morocco.

[5] Masen, 2010, "Enjeux des energies renouvelables, Morrocan Agency for Solar Energy," [Online]. Available: http://www.masen.org.ma. [Accessed 11 4 2016].

[6] A. Ogunjuyigbe, T. Ayodele and C. Monyei, 2015, "An intelligent load manager for PV powered off-grid residential houses," *Energy for Sustainable Development,* vol. 26, pp. 34-42.

[7] S. Faquir, A. Yahyaouy, H. Tairi and J. Sabor, 2016, "Energy management in a hybrid pv/wind/battery system using a type-1 fuzzy logic computer algorithm," *International Journal of Intelligent Engineering Informatics,* vol. 4, no. 3, pp. 229-244.

[8] D. A. Atia, F. H. Fahmey, M. A. Ninet and A. T. Dorrah, 2012, "Modeling and Control PV-Wind Hybrid System Based On Fuzzy Logic Control Technique, pp..," *TELKOMNIKA,* vol. 10, no. 3, pp. 431-441.

[9] P. Garcıa, J. P. Torreglosa, L. M. Fernandez and F. Jurado, 2013, "Optimal energy management system for standalone wind turbine/photovoltaic/hydrogen/ battery hybrid system with supervisory control based on fuzzy logic, international journal of hydrogen energy," *International Journal of Hydrogem Energy,* vol. 38, no. 33, pp. 14146-14158.

[10] M. Hafiane, J. Sabor and H. Gualous, 2015, "Gestion d'énergie d'une Eolienne par une approche mode glissant Application sur le site Cap Sim à Essaouira (Maroc)," in *Congrès international de génie industriel - EducationCM (CIGIMS)*, FEZ.

[11] S. Faquir, A. Yahyahouy, H. Tairi and J. Sabor, 2013, "A comparison between storage methods in a hybrid renewable energy system," in *The Fifth Workshop on Information Technologies and Communication (WOTIC)*, Fes-Morocco.

[12] P. Nema, R. K. Nema and S. Rangnerak, 2009, "A current and future state of art development of hybrid energy system using wind and PV-solar:A Review," *Renewable and Sustainable energy reviews,* vol. 18, no. 8, pp. 2096-2103.

[13] S. Faquir, A. Yahyaouy, H. Tairi and J. Sabor, 2015, "Implementing a Fuzzy Logic Based Algorithm to Predict Solar and and Wind Energies in a Hybrid Renewable Energy System," *International Journal of Fuzzy System Applications,* vol. 4, no. 3, pp. 10-24.

[14] C. Ameur, S. Faquir, A. Yahyaouy and A. Sabri, 2017, "An Approach for revising a fuzzy logic controller using Q-learning algorithm," in *The second International Conference on Intelligent Systems and Computer Vision (ISCV)*, Fes-Morocco.

IOP Conf. Series: Materials Science and Engineering **353** (2018) 012026 doi:10.1088/1757-899X/353/1/012026

MAI statistics estimation and analysis in a DS-CDMA system

A Alami Hassani[1*], M Zouak[2], M Mrabti[3] and F Abdi[2]

[1] Systems & Sustainable Environment Laboratory (SED), Private University of Fez, UPF, Fez, Morocco

[2] Signal Systems & Components Laboratory (SSC), Faculty of Sciences & Techniques, USMBA, Fez, Morocco

[3] Laboratory of Computing and Interdisciplinary Physics (LIPI), Ecole Normale Supérieure, USMBA, Fez, Morocco

E-mail* : alami.hassani@upf.ac.ma

Abstract. A primary limitation of Direct Sequence Code Division Multiple Access DS-CDMA link performance and system capacity is multiple access interference (MAI). To examine the performance of CDMA systems in the presence of MAI, i.e., in a multiuser environment, several works assumed that the interference can be approximated by a Gaussian random variable. In this paper, we first develop a new and simple approach to characterize the MAI in a multiuser system. In addition to statistically quantifying the MAI power, the paper also proposes a statistical model for both variance and mean of the MAI for synchronous and asynchronous CDMA transmission. We show that the MAI probability density function (PDF) is Gaussian for the equal-received-energy case and validate it by computer simulations.

1. Introduction

The Direct Sequence Code Division Multiple Access (DS-CDMA) scheme is robust to frequency selective fading, offers the advantages of interferer diversity and flexibility and has been successfully introduced in commercial cellular mobile communications systems such as (3G, 4G, LTE and LTE-Advanced) as the key technology to improve the spectrum efficiency and flexible user resource allocation [1-3]. However, it suffers from multiple access interference (MAI) caused by the non-orthogonality of spreading codes, particularly for heavily loaded systems [4].

In CDMA systems, direct-sequence spread-spectrum (DS-SS) signals must be power controlled to mitigate multiuser interference access or (MAI), and to properly perform this, signal strengths, or signal "quality," must be estimated. Also, for call handoff between cells in cellular systems, and for rapid performance estimation, signal strength-quality estimation is essential [5] even for more advanced reception techniques such as multiuser detection [6-8] and often require signal quality estimation for proper operation.

Thus, multiple access interference (MAI) is a factor that limits the capacity and performance of DS-CDMA systems [9]. MAI refers to the interference between direct-sequence users. This interference is the result of the random time offsets between signals, which make it impossible to design the code waveforms to be completely orthogonal. While the MAI caused by any one user is generally small, as the number of interferers or their power increases, MAI becomes substantial [10]. Theoretical tools for evaluating the performance in terms of bit error rate (BER) are important in simplifying the system design and deployment tasks. In the recent past, such theoretical evaluations of the BER of various systems have been reported under different conditions.

Single user in AWGN channel was considered in [11] and [12]. Under these conditions, the problem is straightforward and the BER can be represented by the Q-function (or the Gaussian tail probability function) exactly. System performance in AWGN channels considering MAI was addressed in [12-15], where the MAI is modeled as a Gaussian random variable (Gaussian Approximation (GA)). With the GA assumption, the problem was simplified and became tractable with a simple closed form solution. In multi-user multi-path fading conditions, either the GA was used in deriving the average BER [12], [16] or the performance evaluation was based entirely on Monte-Carlo simulations [17].
In this paper, we present an exact, simple and more realistic method to model the statistics of MAI for a DS-CDMA system and DS-BPSK scheme in AWGN channel. Using this model, the characteristic function of the total interference is derived. Simulations are undertaken to prove the analytical work.
The rest of this paper is organized as follows. A description of system model is presented in section 2. In section 3, the statistical model of the MAI in AWGN channel is developed where simulations are used to determine the characteristic function of the MAI.

2. System description

We here consider a DS-CDMA system consisting of K active users who employ coherent BPSK and share a flat fading multiple access channel with AWGN.

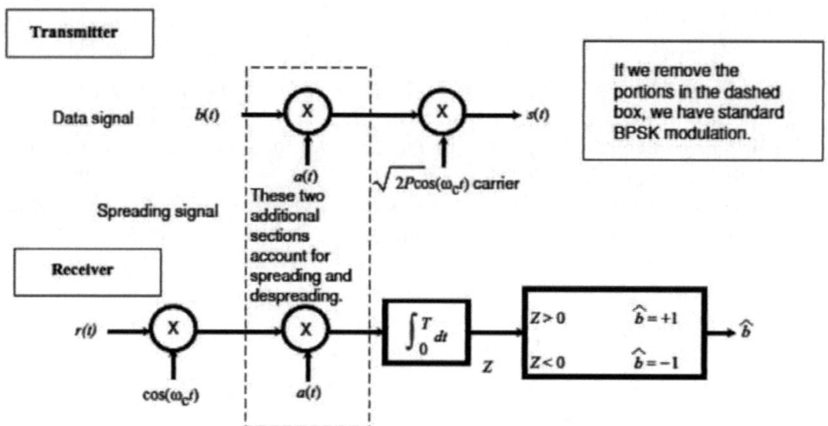

Figure 1. Transmitter and receiver block diagram for DS-CDMA and BPSK modulation

The transmitted signal (spread spectrum signal) of the kth user can be expressed as:

$$s_k(t) = \sqrt{2P_k}\,.\,d_k(t).\,c_k(t)\cos(\omega_0 t) \quad (1)$$

Where P_k denotes the kth user's transmitted signal power, $c_k(t)$ and $d_k(t)$ are the spreading code and data waveforms respectively (we assume rectangular pulses for both), ω_0 is the carrier frequency. Each symbol c_i (usually called a chip) has duration $T_c = T_b/N$, T_b is the bit width and N is the bandwidth expansion factor, or the spreading gain.
In the conventional DS-CDMA detector, and according to [18] the received signal can be modeled as:

$$r(t) = \sum_{k=1}^{K} s_k(t-\tau_k) + n(t) \quad (2)$$

Where n(t) is an AWGN with zero mean and power spectral density $\frac{\eta_0}{2}$, and τ_k is the random phase offset due to propagation, τ_k represents the propagation delay for the Kth user.
The received signal can be written in the form:

$$r(t) = \sum_{k=1}^{K} \sqrt{2P_k}\, d_k(t-\tau_k) c_k(t-\tau_k)\cos(\omega_0 t + \theta_k) + n(t) \quad (3)$$

Where θ_k is the random phase, of user k. τ_k and θ_k are assumed to be independent random variables uniformly distributed respectively over [0,Tb] and [0,2π].

IOP Conf. Series: Materials Science and Engineering **353** (2018) 012026 doi:10.1088/1757-899X/353/1/012026

In the CCR (Fig. 2), the received data sequences are correlated with the signature code to generate correlation functions. Data sequences arriving at the decoder with unmatched signature code result in cross-correlation functions, which in turn, are considered as interference [19].
After a coherent demodulation, the correlated signal is [20]:

$$r_{corr}(t) = r(t)c_1(t)\,cos(\omega_0 t) \quad (4)$$

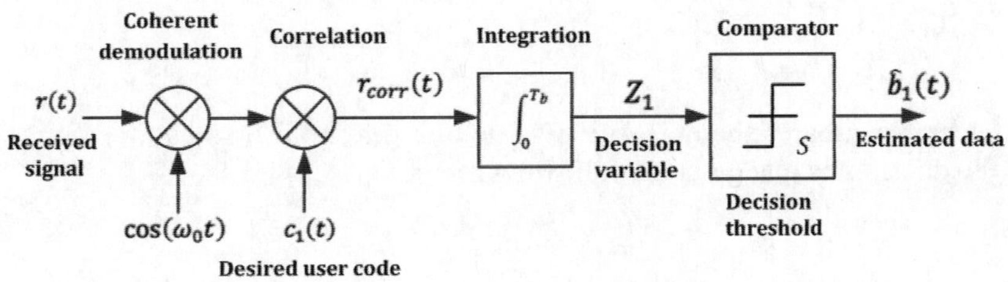

Figure 2. Conventional Correlation Receiver.

Substituting (3) into (4) gives

$$r_{corr}(t) = c_1(t)\,cos(\omega_0 t) \times \left(n(t) + \sum_{k=1}^{K} \sqrt{2P_k}\, d_k(t-\tau_k)c_k(t-\tau_k)\,cos(\omega_0 t+\theta_k)\right) \quad (5)$$

Which can be rewritten as:

$$r_{corr}(t) = c_1(t)\,cos(\omega_0 t) \times \left(n(t) + \sqrt{2P_1}d_1(t-\tau_1)c_1(t-\tau_1)\,cos(\omega_0 t+\theta_1) + \sum_{k=2}^{K} \sqrt{2P_k}\, d_k(t-\tau_k)c_k(t-\tau_k)\,cos(\omega_0 t+\theta_k)\right) \quad (6)$$

Assuming perfect phase coherence, bit timing, and chip timing at the receiver (θ_1 =0 et $\tau_1 = 0$), under a synchronous environment, the correlated signal becomes:

$$r_{corr}(t) = n(t).c_1(t).cos(\omega_0 t) + \sqrt{2P_1}.d_1(t).{c_1}^2(t).cos^2(\omega_0 t) + \sum_{k=2}^{K} \sqrt{2P_k}\, d_k(t-\tau_k)c_k(t-\tau_k)\,c_1(t)\,cos(\omega_0 t)\,cos(\omega_0 t+\theta_k) \quad (7)$$

The DS-CDMA output of the matched filter for the recovery of the user 1 can be expressed as

$${Z_i}^{(1)} = \int_0^{T_b} r_{corr}(t)dt \quad (8)$$

Where ${Z_i}^{(1)}$ is the decision variable for the ith data bit ${d_i}^{(1)}$for user n° 1.

The resulting decision variable, as shown below, comprises three components, namely, the AWGN V, the desired information bit A, and the MAI term I which will be analyzed in more detail in the section below.

IOP Conf. Series: Materials Science and Engineering **353** (2018) 012026 doi:10.1088/1757-899X/353/1/012026

$$Z_i^{(1)} = \int_0^{T_b} n(t).c_1(t).cos(\omega_0 t)\,dt + \int_0^{T_b} \sqrt{2P_1}.d_i^{(1)}(t).c_1^2(t).cos^2(\omega_0 t)\,dt + \int_0^{T_b}\sum_{k=2}^{K}\sqrt{2P_k}.d_i^{(k)}(t-\tau_k).c_k(t-\tau_k).c_1(t).cos(\omega_0 t)\,cos(\omega_0 t+\theta_k)dt = V + A + I \quad (9)$$

The threshold detector compares the correlation value to a threshold S so as to extract the transmitted data [21]. Here the decision is made via the following relation:

$$\begin{cases} Z_i^{(1)} \geq S \Rightarrow \hat{b}_i^{(1)} = 1 \\ Z_i^{(1)} < S \Rightarrow \hat{b}_i^{(1)} = -1 \end{cases} \quad (10)$$

3. Statistical modelling of MAI

This section will focus on the properties of the MAI, particularly the statistical properties. But first, the general expression of the MAI and its power should be developed.

3.1. MAI analysis

The total MAI incurred by the reference user is given by the sum of individual interferer MAI contributions as:

$$I = \sum_{k=2}^{K} I_k^{(1)} \quad (11)$$

The expression for the individual interferer MAI contributions were given in equation (9):

$$I = \int_0^{T_b}\sum_{k=2}^{K}\sqrt{2P_k}.d_i^{(k)}(t-\tau_k)c_k(t-\tau_k)\,c_1(t)\,cos(\omega_0 t)\,cos(\omega_0 t+\theta_k)\,dt \quad (12)$$

this expression can be written as:

$$I = \int_0^{T_b}\left(\sum_{k=2}^{K}\sqrt{\frac{P_k}{2}}.d_i^{(k)}(t-\tau_k).c_k(t-\tau_k).c_1(t).cos(\varphi_k)\right)dt = \sum_{k=2}^{K}\sqrt{\frac{P_k}{2}}\int_0^{T_b} d_i^{(k)}(t-\tau_k).c_k(t-\tau_k).c_1(t).cos(\varphi_k)\,dt \quad (13)$$

where $\varphi_k = \theta_k - \omega_0\tau_k$

τ_k and θ_k are modelised as independent random variables uniformly distributed respectively over [0,Tb] and [0,2π] so $\cos(\varphi_k)$ is independent from spreading code and data. $E[\cos(\varphi_k)] = 0$, and $E[\cos^2(\varphi_k)] = \frac{1}{2}$. The MAI is therefore a zero-mean random variable with $E[I] = 0$ and the variance of this term of interference can be shown to be:

$$var[I] = E[(I)^2]$$

IOP Conf. Series: Materials Science and Engineering **353** (2018) 012026 doi:10.1088/1757-899X/353/1/012026

$$= E\left\{\left(\sum_{k=2}^{K}\sqrt{\frac{P_k}{2}}\int_0^{T_b} d_i^{(k)}(t-\tau_k).c_k(t-\tau_k).c_1(t).cos(\varphi_k)\,dt\right)^2\right\} \quad (14)$$

With

$$I_k^{(1)^2} = \frac{P_k}{4}.E\left[\left(\int_0^{T_b} d_i^{(k)}(t-\tau_k).c_k(t-\tau_k).c_1(t)dt\right)^2\right] \quad (15)$$

$I_k^{(1)^2}$ is the variance of the MAI contribution from user k on the reference user 1.

To model $E[(I)^2]$, several expressions have been proposed [22,23], all agree that it's proportional to the power P and the square of symbol duration T_b. Furthermore, for the same period sequences N, $I_k^{2(1)}$ is expressed as [22]

$$I_k^{2(1)} = \frac{NT_c^2}{6}P_k \quad (16)$$

Thus $$E[(I)^2] = \frac{NT_c^2}{6}\sum_{k=2}^{K} P_k \quad (17)$$

Assuming that all users exhibit equal power $P_k = P_{moy}$, then:

$$E[(I)^2] = \frac{NT_c^2 P_{moy}}{6}(K-1) \quad (18)$$

3.2. Simulation results

To determine the MAI properties (mean and variance), theoretical analysis are compared to simulation results witch have been performed for both synchronous and asynchronous DS-CDMA systems.

3.2.1. MAI mean. We first determined, by simulation, MAI mean versus the number of users in both synchronous and asynchronous cases. For the asynchronous case, offsets between active users and desired user codes are generated randomly. The results are shown on the following Fig. 3.

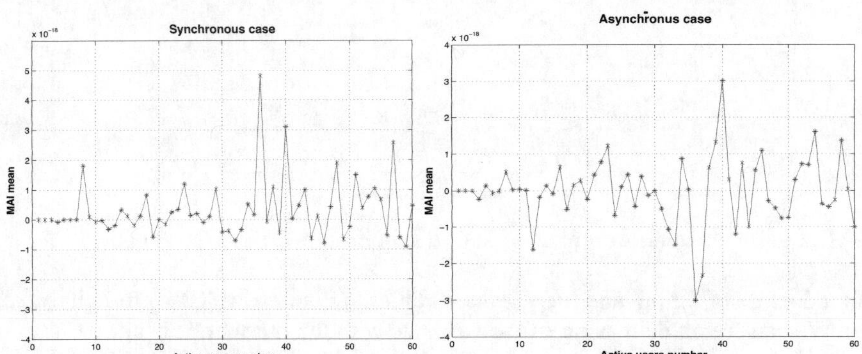

Figure 3. MAI mean versus number of active users

The means are viewed as a random process around the value null in both the synchronous and asynchronous cases. These results are obtained whatever the value of N (63, 127, 255) is.

To confirm this result, we undertake a statistical study in which we determine the frequency of appearance of the values obtained for different numbers of active users, for different lengths of the spreading sequence, for different powers (same powers for all users) and this in both the synchronous and asynchronous cases. Examples of results are presented in the graphs on Fig. 4 below.

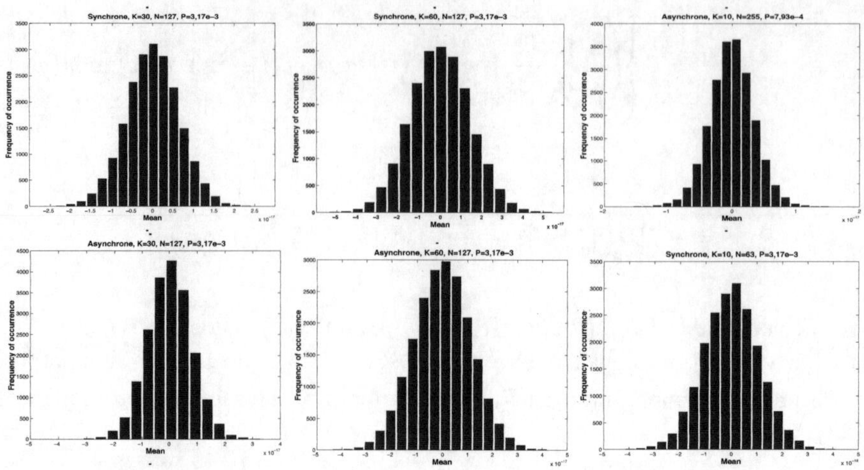

Figure 4. MAI mean distribution for different values of P, N and K in synchronous and asynchronous cases

In all cases, distributions show Gaussian shape with '0' as average value. Thus this result highlight that the MAI is Gaussian distribution and can be considered as an additive Gaussian noise, which validates the approximation adopted by several authors. The interference terms have zero mean and only their variances should be considered.

3.2.2. MAI variance: synchronous case. Since the variance is depending on the number of users, we have simulated this variance according to K. The result is shown in Fig. 5.

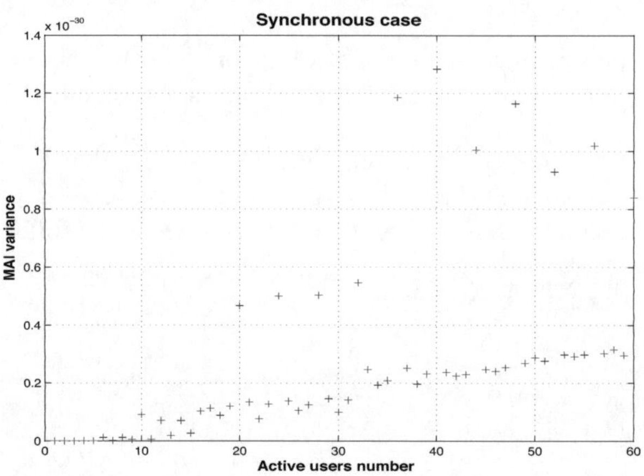

Figure 5. MAI variance versus number of active synchronous users, N=63 and P=3.17 10^{-3}(ua)

We observe a random distribution, and the variance values are very low (10^{-30}) for a user average power 3.17 10^{-3} (AU). We can see an increasing of the variance with the number of users.
To determine the type of variance, the same statistical study made previously for MAI mean is made for the MAI variance. The graphs in Fig. 6 show some examples of the simulations results.

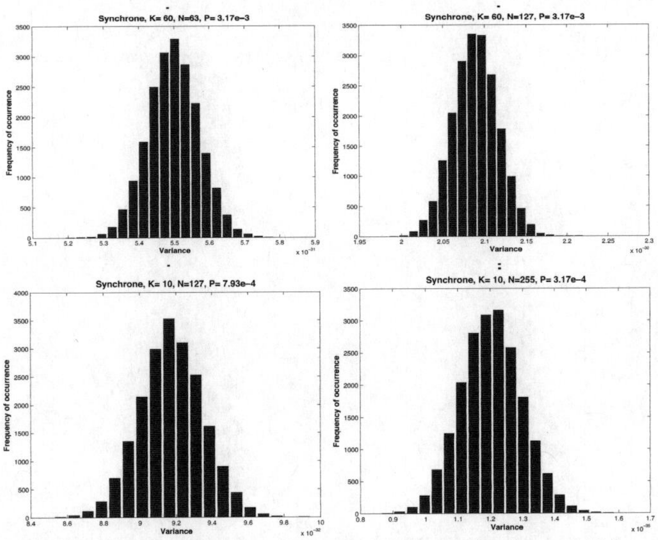

Figure 6. MAI variance distribution for different values of P, N and K in synchronous case

We infer that the distributions are Gaussian, but the values are very low. Indeed, in a synchronous case, bit of each user's data stream are temporally aligned at the receiver and thus the spreading sequences retain their desired low cross-correlation properties. As a result, the MAI incurred upon reception is minimal (if not totally prevented), hence the very negligible obtained values.

3.2.3. MAI variance: asynchronous case. In the asynchronous case, spreading codes of different users are shifted randomly compared with the desired user one. The MAI variance versus the active users number is shown in Fig. 7.

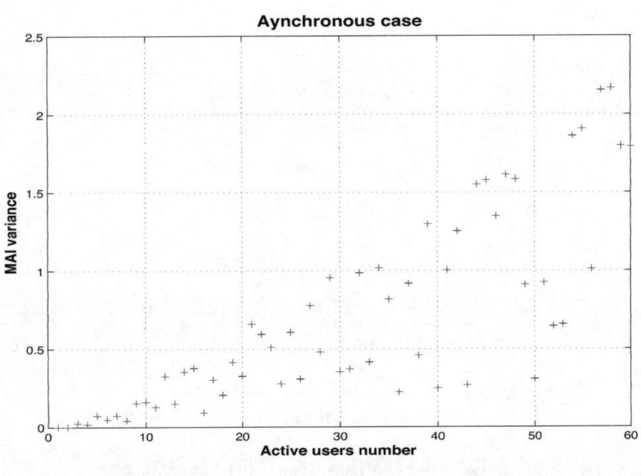

Figure 7. MAI variance versus number of active asynchronous users, N=63 and P=3.17 10^{-3}(ua)

We notice the random behavior the evolution. The MAI variance increases when the number of users increases. These variance values are not negligible compared to the user power value and to the variance in the synchronous case. We will then determine the MAI average variance using a statistical study. The results are shown on graphs in Fig. 8.

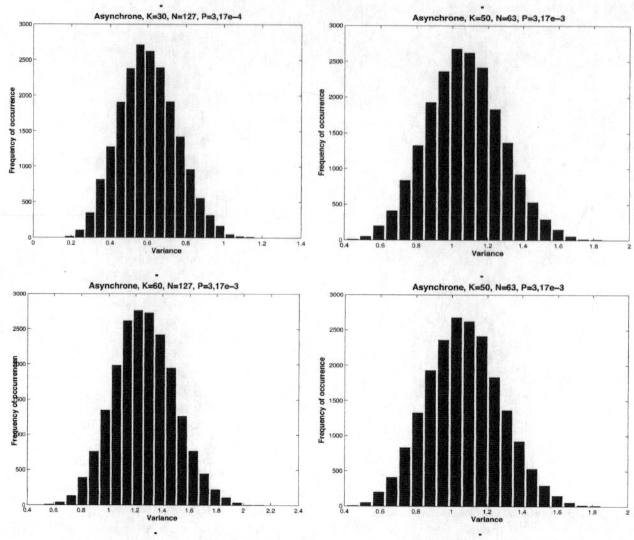

Figure 8. MAI variance distribution in the asynchronous case

The graphs show MAI variances histograms performed for different sequences lengths, different powers and different active users numbers. Distributions are Gaussian with large values compared to the users power.

Using these simulation results, we can plot the MAI variance (average values of the Gaussians) versus active users number or users power.

The MAI variance versus average users power for different values of K, Tc and N are shown in Fig. 9.

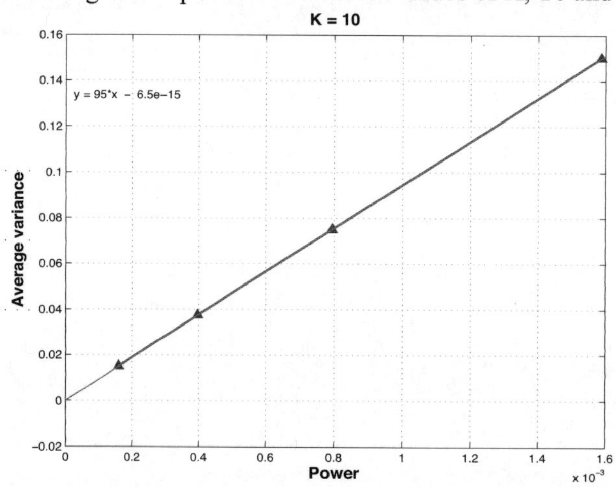

Figure 9. MAI variance versus average users power (in Watt) for K=10, N=63, Tc=1 (unit) in the asynchronous case

We note a perfectly linear behavior, the data adjustment provides a slope of 95. The theoretical slope (Eq 18) $\frac{NT_c^2}{6}(K-1)$= 94.5 is in good agreement with the simulation results. Other simulations with different parameters values confirm this result.

Fig. 10 shows an example of the MAI variance versus active users number K plot for N=127, Tc=0.5 and average power P_{moy}=3.17*10^{-3}(W). All obtained results for different values of N, Tc and P_{moy} are qualitatively the same.

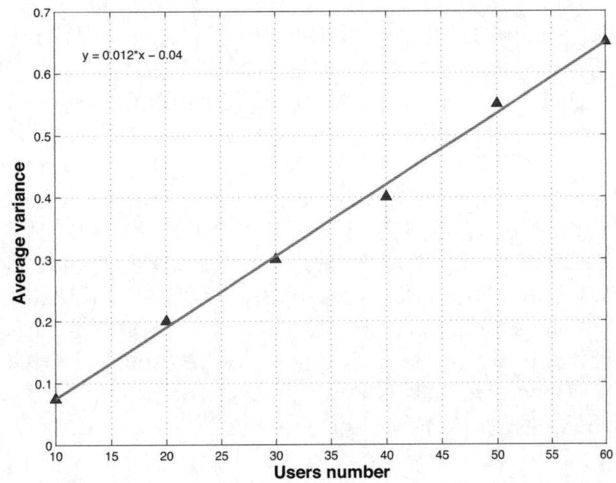

Figure 10. MAI variance versus active users number K for $P_{moy} = 3.17.10^{-3}(W)$, N=127 and Tc=0.5 in the asynchronous case

We notice the perfectly linear behavior with a slope of 0.012. The theoretical slope (Eq 18) $\frac{NT_c^2 P_{moy}}{6} = 0.016$ is also in good agreement with the simulation result.

4. Conclusion

Through this study, we developed an analytical model of the MAI, we showed that MAI can be considered as a noise with zero mean and variance expressed:

$$\begin{aligned} E[(I)^2] &= \frac{NP_{moy}T_c^2}{6}(K-1) \\ &= \frac{P_{moy}T_b^2}{6N}(K-1) \end{aligned} \quad (19)$$

The simulation study confirms these results.

The MAI variance allows the contribution estimation of interference (MAI) in the signal-to-noise-plus-interference ratio (SNIR) in the CDMA system. So In a future work, we will use it to present performance analysis of CCR by evaluating the instantaneous SNIR and the BER.

5. References

[1] ETSI TS136.300 (V13.2.0) 2016 LTE, Evolved Universal Terrestrial Radio Access (EUTRA) and Evolved Universal Terrestrial Radio Access Network (EUTRAN). Overall description.
[2] ETSI TR 136 913 (V9.0.0) 2010 LTE, Requirements for further advancements for E-UTRA (LTE-Advanced). Overall description.
[3] 3GPP TR36.814 (V9.0.0) 2010 Technical Specification Group Radio Access Network; Evolved Universal Terrestrial Radio Access (E-UTRA), Further advancements for E- UTRA physical layer aspects. Overall description.
[4] S.Pradhana, S.Chattopadhyayb, S.Rautc 2014 A Novel Orthogonal Minimum Correlation Spreading Code in CDMA System. International Journal of Wireless and Microwave Technologies.
[5] V. K. Garg 2000 IS-95 CDMA and cdma2000. Cellular/PCS Systems Implementation. Prentice-Hall,Upper Saddle River, NJ.
[6] S.Verdu 1998 Multiuser Defection. New York Cambridge Univ.
[7] P. Patel, J. Holtzman 1994 Analysis of a Simple Successive Interference Cancellation Scheme in a DS-CDMA System. IEEE JSAC, vol. 12, no. 5, pp. 796-807.

[8] D. Divsalar, M. K. Simon, D. Raphaeli 1998 Improved Parallel Interference Cancellation for CDMA. IEEE Trans. Comm., vol. 46, no. 2, pp. 258-268.

[9] Shikant J.Honade, Dr.P.V.Ingole 2012 Removal of Multiple Access Interference in DS-CDMA System. International Journal of Scientific and Reserch Publications, Volume 2, Issue 6.

[10] Mohamed H. Essai, Abdelrahim H. Gadelkarim 2017 Performance of Linear Multiuser Detection in Gaussian and non-Gaussian Channels. Journal of Multidisciplinary Engineering Science and Technology (JMEST) ISSN: 2458-9403 Vol. 4 Issue 3.

[11] J. A. Ney da Silva and M. L. R. de Campos 2003 Performance Comparison of Binary and Quaternary UWB Modulation Schemes. in Proc. IEEE GLOBECOM '03, pp. 789-793.

[12] J. Zhang, T. D. Abhayapala, and R. A. Kennedy 2003 Performance of ultra- Wideband Correlator Receiver using Gaussian Monocycles. in Proc. IEEE ICC '03, pp. 2192-2196.

[13] G. Durisi and S. Benedetto 2003 Performance Evaluation and Comparison of Different Modulation Schemes for UWB Multiaccess Systems. in Proc. IEEE ICC'03, pp. 2187-2191.

[14] F. Ramirez-Mireles 2001 Performance of ultrawideband SSMA using time hopping and M-ary PPM. IEEE J. Select. Areas Commun., vol. 19, pp. 1186-1196.

[15] M. Z. Win and R. A. Scholtz 2000 Ultra-wide bandwidth time-hopping spread-spectrum impulse radio for wireless multiple-access communications. IEEE Trans. Commun, vol. 48, pp. 679-689.

[16] V. Venkatesan and al 2003 Performance of an Optimally Spaced PPM Ultra- Wideband System with Direct Sequence Spreading for Multiple Access. in Proc. IEEE VTC'03-Fall , pp. 602-606.

[17] S. Niranjayan, A. Nallanathan, and B. Kannan 2004 Delay Tuning Based Transmit Diversity Scheme for TH-PPM UWB: Performance with RAKE Reception and Comparison with Multi RX Schemes. in Proc. Joint UWBST and IWUWBS'04.

[18] Adit Kurniawan 2014 Power-Controlled Multicarrier and Multiuser Detection Techniques in Wireless CDMA Systems. International Journal on Electrical Engineering and Informatics (IJEEI), v 6, No 3.

[19] A. J Viterbi 1995 CDMA: Principles of Spread Spectrum Communication. Reading MA: Addison-Wesley.

[20] Ali M. Fadhil, Haider M. AlSabbagh 2002 Performance Analysis for Bit-Error-Rate of DS-CDMA Sensor Network Systems with Source Coding. TELKOMNIKA,v.10, No.1, p. 165-170.

[21] A.E. El-Mahdy, M.Ahmed 2013 Low Complexity Selective Adaptive Multicarrier DSCDMA Receiver. TELKOMNIKA, v.11, No.2, p. 321-330.

[22] Rappaport, T. S 2002 Wireless Communications: Principles and practice 2nd edition. Prentice-Hall.

[23] Chan, ChiKin, Lam and WongHing 1994 Conference Recordc, International Conference On Communications (IEEE) v. 3, p. 1516-1520.

Institute of Physics
Dirac House, Temple Back
Bristol BS1 6BE UK

ISSN: 1757-8981
ISBN 978-1-5108-6311-8
9 781510 863118

CURRAN ASSOCIATES INC.
proceedings.com